Exploration and Mapping
of the
American West
Selected Essays

Edited by
Donna P. Koepp

Occasional Paper No. 1

Map and Geography Round Table
of the
American Library Association

SPECULUM ORBIS PRESS
CHICAGO
1986

Published by
Speculum Orbis Press
207 W. Superior
Chicago, Illinois
60610 USA
for the
Map and Geography Round Table
of the
American Library Association

Typesetting supplied by
Wesley A. Brown
and
Bowne of Denver

CONTENTS

FOREWORD

On the pages that follow, a story unfolds that provides us with a glimpse into the past when the American West was little known and almost totally unmapped. A continuous thread of cartographic history runs throughout these papers, resulting in an examination of the primary product of map librarianship, the map. The eight authors who have contributed to this work share with the reader the story of how the land was explored, measured and mapped, and how early maps and cartographers influenced one another, in the intriguing political climate that existed throughout this period of exploration.

This work marks the beginning of an occasional paper series by the Map and Geograpy Round Table of the American Library Association. It combines the talents and perspectives of a wide range of individuals all of whom contributed the papers contained herein at programs devoted to the mapping of the trans-Mississippi West at annual meetings of MAGERT. The interest and excitement generated by these programs encouraged MAGERT to share these papers with others beyond the map library community.

This volume is presented for the enlightenment and pleasure of all who appreciate the beauty and history of the map; for the historian, the geographer, the map collector, and the map librarian.

It is due to the generosity of two donors that this volume has become a reality. Wesley A. Brown of Denver, was the first to express a desire to assist with the publication. Through his efforts, typesetting and layout was supplied by Bowne of Denver. John T. Monckton, of the John T. Monckton Gallery Ltd. in Chicago, pledged the necessary amount to complete the publication of the work by his Speculum Orbis Press. The membership of the Map and Geography Round Table joins me in expressing our sincere gratitude.

I would also like to thank members of the Publications Committee who helped through the years of editing and compiling this publication and continued to have faith that it would someday actually be published. A special thank you to David Cobb who is always there to answer questions, make and facilitate contacts and to lend encouragement whenever it is needed.

DONNA P. KOEPP
University of Kansas

iv

ILLUSTRATIONS

MAPPING THE TRANS-MISSISSIPPI WEST: ANNOTATED SELECTIONS

Kenneth Nebenzahl

Kenneth Nebenzahl, Inc.
Chicago

Originally presented in 1981 to the American Library Association in San Francisco. Revised and edited by David A. Cobb, Map & Geography Librarian, University of Illinois at Urbana-Champaign.

*T*here are two key words in an accurate physical description of the American West: vastness and complexity. The distance from the Mississippi River to the Pacific Ocean is approximately sixteen hundred and fifty miles; this results in an area of over two million square miles — larger than all of Europe, excluding only Russia. Within this area there are great contrasts: complexes of mountain ranges; drainage systems; rich, fertile regions; splendid forests; vast wastelands; deserts; canyons; and extremes of temperature and moisture. Moreover, it is three miles in elevation from top to bottom.

All of this remained unknown to Europeans until, quite by accident, as with so many discoveries, the second quarter of the sixteenth century saw the beginning of the region's exposure to explorers from afar. It was still to be over three centuries later before all the secrets of the geography of the American West were to become known, and mapping would be possible of the kind we are used to seeing in modern times, on which the blanks have been filled.

In 1528, Alvar Núñez Cabeza de Vaca was shipwrecked on the Texas Coast. First a prisoner of Indians, then, after his escape, a medicine man himself, Cabeza de Vaca was the first European to penetrate what is now the western United States. He wandered during eight years to Mexico, by way of New Mexico, and lived to publish his account. Núñez Cabeza de Vaca arrived in Mexico with tales of fabulous cities and enormous wealth of the Indians of the North. This stimulated great interest and triggered a series of explorations.

This survey is designed to give a brief view of some of the major maps from the 1540's to the 1840's: the maps that reported progress in knowledge of the area, as opposed to those that were retrogressive, or which followed the many misconceptions and aberrations that figured in the region's cartographic development — although we must list a few of these as well.

Considering the nature of the region, it is understandable that early geographers might have created curious, imaginary and apocryphal features. The most persistent of these were:

1) the elongated westward bulge of the continent
2) the central "height of land" from which point many, if not all, of the major rivers had their source
3) the "Island of California"
4) the "Sea of the West"
5) the great east to west rivers — flowing from the central highlands to the Pacific directly across the Great Basin, Sierra Nevada, and the Coastal Range; while it seemed to take forever to organize the Mississippi, the Missouri, the Colorado, the Columbia and their tributaries.

Printed maps, rather than manuscripts, have generally been used for this paper since they were the documents that circulated widely and influenced people in many places.

1) "Novae Insvlae, XVII Nova Tabvla." (In Ptolmaeus, Claudius. *Geographia Universalis*... Basel: Sebastian Munster, 1540.)

This is the first separate map of America and the first to show the New World as an insular land mass in addition to showing continuity between North and South America. Not surprisingly, it has no place names in the West as Munster published it the same year Coronado set out to discover Quivira. The false Sea of Verrazano is shown quite dramatically as is Magellan's ship Victoria sailing in the Pacific after emerging from the Strait, later named for the explorer.

2) "Vniversale." Venice: Giacomo Gastaldi, 1546.

The center of European map-making at this time was Venice and its most important cartographer was Gastaldi. This map was perhaps his greatest work. It was for sale within a half-century after Columbus's voyages and it shows exceptional progress in understanding the distant regions, but note that Asia is shown joined to America. It does show Niza place names appearing in the Southwest: Le Sete Cita, Cipola, and Tabursa.

3) "Universale Della Parte del Mundo Nuovamente Ritrovata." Giacomo Gastaldi.
(In Ramusio, Giovanni. *Della navigationi e Viaggi.* Venice, 1556.)

Ramusio was a Venetian publisher who compiled a great collection of voyages and Gastaldi made a number of the maps which accompanies this work. This map of America dates from just ten years after his world map. Numerous Coronado place names make their first appearance on a map: Quivira, Cicuich, Axa, Cucho and Tiguas; along with the Civola of Niza.

4) "Il Desegno del Discoperto Della Noua Franza." Bolognino Zaltieri, 1566.
(In *Geographia Tavole Moderne di Geographia de la Maggior Parte del Mondo,* Rome, 1566.)

This is the first separate map devoted to North America to have been published. Zaltieri worked closely with Gastaldi and was also one of the leading sixteenth century Italian map-makers. It is the first map to show the Strait of Anian dividing Asia from North America. The maps of the sixteenth century joining Asia with North America, or showing a Strait of Anian, are not for that reason alone, particularly significant — after all, no European had been to this remote area. The entire American West is dominated by the designation "Quivira" and the complex topography had not yet begun to be unraveled.

5) "Nova et Avcta Orbis Terrae Descriptio ad Vsvm Nauigantium Emendate Accomodata." Duisberg: Gerard Mercator, 1569.

One of the most influential maps of its era, Mercator's map dominated geographical thinking for many decades, even after discoveries had revealed its errors. It presented the isogonic cylindrical projection, soon to become known as the Mercator projection, for the first time and helped liberate cartography from classical Ptolemaic thought. This map was the source for the following one — both its new information and its distortions.

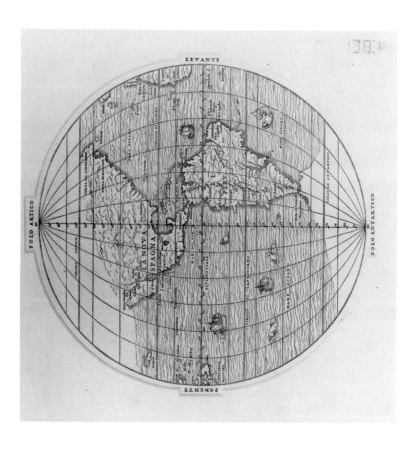

Fig. 1. "Universale Della Parte del Mundo Nuovamente Ritrovata." Giacomo Gastaldi.

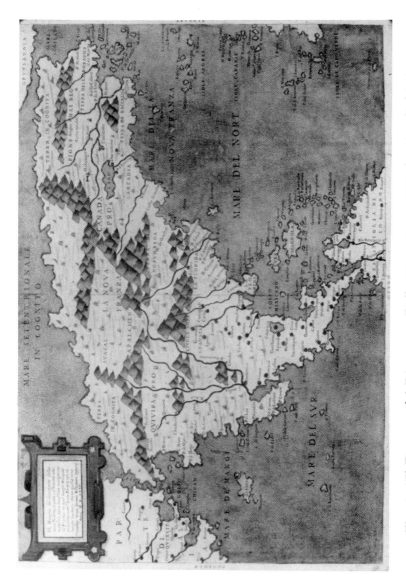

Fig. 2. "Il Desegno del Discoperto Della Noua Franza," Bolognino Zaltieri.

6) "Americae Sive Novi Orbis, Nova Descriptio." (In *Theatrum Orbis Terrarum*. Antwerp: Abraham Ortelius, 1570.)

This is the first map of America to be published in a modern atlas. His delineation of America is derived from Mercator's great map of the previous year and other sources. Unlike many other mapmakers, Ortelius scrupulously credited the cartographers whose maps he used. His lists of these sources are of great bibliographical value.

He popularized the erroneous concept of a North American continent bulging westward to a great width which was adopted from Mercator. This map shows enormous rivers traversing the continent east and west, but no Mississippi or equivalent. Ortelius's maps were very widely circulated and influential; they were the best selling atlases that had so far appeared.

7) "Novvs Orbis." Francis Gaulle. (In *De Orbe Novo*. Paris: Pietro Aughiera, 1587.)

This is the first appearance of the name New Mexico on a published map and is the earliest map known to show the New Mexico lake reported by Espejo on his expedition to the Hopi villages. It appeared in a most important Elizabethan compilation issued by Richard Hakluyt, known as the greatest propagandist for the British empire during the reign of Elizabeth I.

8) "Maris Pacifici (quod vulgò Mar del Zur), cum Regionibus Circumiacentibus . . ."
(In *Theatrum Orbis Terrarum*. Antwerp: Abraham Ortelius, 1589.)

In 1579 Francis Drake landed on the California coast, took possession of the area, and named it New Albion. It is from this voyage that England later based her claims on the Northwest Coast. Drake actually caused more controversy than he resolved, especially as far as the cartography of the California coast was concerned. His round the world voyage (1577-80) took place between the time of Ortelius's map of 1570 and this map of 1589.

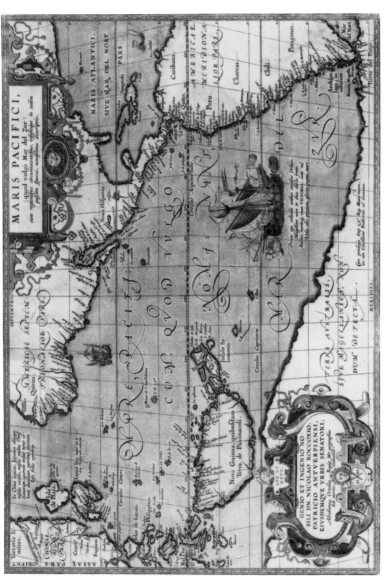

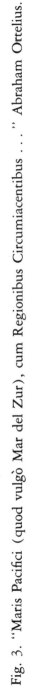

Fig. 3. "Maris Pacifici (quod vulgò Mar del Zur), cum Regionibus Circumiacentibus . . ." Abraham Ortelius.

Ortelius dropped Niza and the Coronado place names from this map for some unknown reason. The sources for the names he has in western North America, however, remain unknown. They are not from the Mercator map, which has very few place names on the West Coast; but the overall shape is derived from Mercator.

9) "Granata Nova et California." Cornelius van Wytfliet. (In *Descriptionis Ptolemaicae Augmentum.* Louvain, 1597.)

Based on Mercator-Ortelius information this is the first separately published map of the California region. It shows the large New Mexico lake and the legendary seven cities around it. Wytfliet published the first atlas devoted exclusively to America and this was one of the nineteen maps it contained.

10) "The North Part of North America." Henry Briggs. (In Purchas, Samuel: *Purchas his Pilgrimes.* London, 1625.)

The eminent Henry Briggs published an account of North America in a successor collection of voyages to that of Hakluyt, published by Samuel Purchas in 1625. This map, by Briggs, accompanies the narrative. While it contains considerable information about Brigg's friends and associates in the Arctic area, derived no doubt from Luke Fox, whose patron Briggs was, this is an extremely important map for the West Coast since it contains data from the Vizcaino expedition of 1602-03.

Sebastian Vizcaino led the first scientific voyage up the Pacific Coast of California, seeking to find ports for the Manila to Acapulco galleons as a stop-over. He charted the coast as far north as Cape Blanco in Oregon. His information was used for the next 200 years and many of his place names still exist. Did Vizcaino provide Briggs with the electrifying information that California was an island, a concept that this map first helped to popularize? Not directly. The Dutch captured a Spanish map at sea, and this information, or should I call it misinformation, came into Brigg's hands from Amsterdam. Publication of this map is considered to have been the

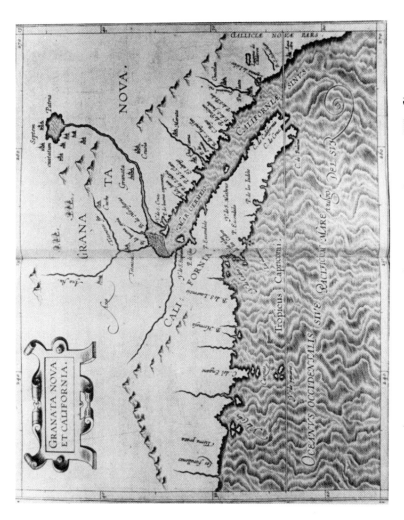

Fig. 4. "Granata Nova et California." Cornelius van Wytfliet.

Fig. 5. "The North Part of North America," Henry Briggs.

launching of the popular idea of California as an island which continued, as we shall see, for 100 years.

11) "America Septentrionalis." 1640.
(In Jansson, Jan: *Novus atlas absolutissimus* . . . .
Amsterdam, 1658.)

This is one of the most attractive maps of North America and is an excellent example from the golden age of Dutch cartography. Unlike the "plain" surveys of the nineteenth century explorers Jansson relied heavily on decoration. Carl Wheat opined that the best to be said for this map is that the animals are good!

It typifies the misinformation regarding: the Rio Grande, the New Mexico lake, and of course, most dramatically, the insular California.

12) "Ameriqve Septentrionale." Nicolas Sanson. 1650.

Sanson was the first to break the clichés of imaginary cartography established by Mercator and Ortelius in the previous century, and followed by Blaeu, Hondius, Jansson and others. This map provides new factual information: this is the first map with Santa Fe and other "real" place names; we also see for the first time the presence of all five Great Lakes. Sanson must have seen a new map, but thus far his sources remain unknown. For all its cartographic progress the Rio Grande continues to be geographically inaccurate.

13) "Le Nouveau Mexique Appellé Aussi Nouvelle Grenade et Marata. Avec Partie de Californie." Paris: Vincenzo Coronelli, [1685].

The best delineation of New Mexico to date is shown on this map. Coronelli's source was the Comte de Peñalosa, the former Governor of the Province of New Mexico, who had recently defected to the French. This became the prototype map of the region, adopted by many of the European map publishers, and used for decades. A large number of place names appear in the west and Santa Fe is correctly located on the east side of the river.

14) "America Settentrionale colle Nuoue Scoperte."
Vincenzo Coronelli.
(In *Atlante veneto.* Venice, 1695-97.)

Similar to his 1685 map in detail this map covers only the area west of the Great Lakes/Mississippi region. Again he incorporates and credits Penalosa for his information. As important as this map is, it shows the Mississippi far to the west with its mouth located in southern Texas just north of the Rio Grande.

15) "Carte du Mexique et de la Floride."
Paris: Guillaume Delisle, 1703.

Now we begin another new era — which coincides with a new century. You will become familiar with this name, Delisle, and, that is as it should be, for this family dominated cartography during the entire first half of the eighteenth century. This map is considered a "towering landmark for Western cartography" and rightfully so.

Among a collection of advancements, this is: the first printed map with the name of the Colorado River, and to have it nearly geographically correct; and the first map to accurately delineate the mouth of the Mississippi and much of the Mississippi Valley. It is interesting to note that, for all his progress, Delisle still wavers on California as an island.

16) "Carte du Canada ou de la Nouvelle France."
Paris: Guillaume Delisle, 1703.

Here Delisle, although one of the great cartographers of his time, in a map important for its delineation of the Great Lakes and Canadian Arctic regions, falls for the aberration of the "Riviere Longue" promoted by Baron Lahontan. The Munchausian explorer, whose own account was published this same year, retarded the correct understanding of the Lower Missouri River and the whole area west of the Upper Mississippi for decades.

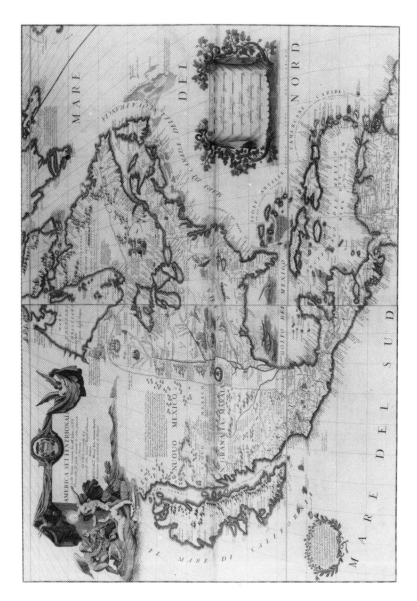

Fig. 6. "America Settentrionale colle Nuoue Scoperte." Vincenzo Coronelli.

17) "Passage par Terre a la California." Eusebio Kino. 1701. (In Jesuits, Letters from Missions: *Lettres edifiantes....* Paris, 1705.)

Father Kino includes his overland route to California which enabled him finally to dispel the theory of California as an island. Father Kino's important discovery was to be popularized by Guillaume Delisle on a map he published in 1722 which we shall soon discuss. This small map is a great monument in the unfolding of the mysteries of the West.

18) **"Carte de la Louisiane et du cours du Mississippi." Paris: Guillaume Delisle, 1718.**

This map finally shows the Missouri River system with some accuracy and it became a source map for delineating the Mississippi. Also, importantly, St. Denis' route from Louisiana into the Spanish Southwest is indicated, reflecting actual French observation of the area for the first time.

19) **"Carte d'Amerique." Paris: Guillaume Delisle, 1722.**

Here is the map on which Father Kino's information about California was first broadcast to the world. Thus ended the 100 year California as an island era, although many European map publishers continued for another whole generation to show the island, California, rather than invest in revising their copperplates or having new ones engraved.

20) **"Carte des Nouvelles Découvertes au Nord de la Mer du Sud." 1752. (In Buache, Phillippe. *Considérations géographiques et phisiques....* Paris, 1781.)**

Although I promised to show progress, instead of retrogression, this map characterizes — caricatures if you will — the utter state of confusion into which the European mapmakers had fallen at mid-eighteenth century regarding the American West.

Let us ignore the troubles with Alaska and the Strait of Juan de Fuca and the author's desperate attempt to delineate a northwest passage to Europe. Rather, we should concentrate on this first appearance of the Great Sea of the West; this monstrous aberration, to say the least, arrested progress for some time. While the producers of this map were considered to be in the vanguard of scientific geographical knowledge, ignorance triumphed in this particular instance.

21) "Carta Reducida del Oceano Asiático, ó Mar del Súr." Madrid: Miguel Costansó, 1771.

Miguel de Costansó was a Spanish cosmographer and Army engineer. His accounts of the expedition from Mexico to settle Alta California in 1769 are a major contribution to California history. This is the superlatively rare printed version of Costansó's manuscript map of 1770. In its published form it continued to be the basis for California maps for a long time to come. The place names are derived in many cases from as early as the account of Vizcaino, in addition to those bestowed by this expedition.

This important survey of the California coast north to Monterey was made under Governor Portolá after the Spanish decision to occupy Upper California. Included on this expedition were Fathers Serra and Crespi, known for planting the faith in California.

22) "Plano Géographico, de la Tierra Descubierta, Nuevamente, a los Rumbos Norte, Noroeste y Oeste, del Nueva Mexico . . . Ano 1778." Don Bernardo de Miera y Pacheco.

This is the first map ever made of the central Rocky Mountains and Great Salt Lake region. In 1776 two Franciscan Fathers, Dominguez and Escalante, set out from Santa Fe with the permission of the Governor of New Mexico to attempt to find a direct overland route from Santa Fe to Monterey, California by way of the higher latitudes, thus avoiding the difficulty of crossing the Grand Canyon of the Colorado River and desert regions to the west. They moved north from Santa Fe, passed through western Colorado, and arrived at the Green River which they called Rio San Buenaventura. They proceeded westward and passed south of the Uintahs, crossed the Wasatch and came to Utah Lake. So far as we know they were the first white men to look upon its waters. At this point, exhausted and short of provisions, they decided to return to Santa Fe. They passed through southern Utah, crossing the Grand Canyon, again, presumably the first white men to have done so, and then turned east for the trek back to New Mexico.

While the expedition failed in its principal original objective, it did constitute the most important exploration achieved before the nineteenth century in the Southwest. It was the farthest advance of Spaniards into the interior of North America since Coronado; and like Coronado it was not followed up. It was to be a half century later that fur hunters and other adventurers from the U.S. would make their way to modern Utah.

Meanwhile, Don Bernardo de Miera y Pacheco made this important manuscript map of the areas traversed. It gives a great deal of topographical information and it is obviously based on information obtained from Indians and discoveries of earlier Spanish expeditions as well as their own firsthand knowledge gained during the journey of 1776. The legends are of particular interest. The first points out that the Comanche now dominated the buffalo plains as far as the Texan border and after continual warfare had finally displaced the Apache. Another reference is to the Grand Canyon, where the Colorado "flows between walls of red rock very deep and rugged."

23) "Carte Générale du Royaume de la Nouvelle Espagne. . ." Alexander von Humboldt, 1804.
(In Humboldt, Alexander: *Atlas Géographique et Physique du Royaume de la Nouvelle-Espagne,* Paris, 1811.)

This is the best map of the American West published to date. It was compiled in Mexico City in 1803 using all available sources including the rare cooperation of the Mexican authorities. It exemplifies the end of the Spanish mapping of the American Southwest, or the mapping using Spanish sources, and ends that era on a very accurate note with a great monument in the history of cartography.

24) "A Map of Lewis and Clark's Track, Across the Western Portion of North America from the Mississippi to the Pacific Ocean . . ." 1804,5, & 6.
(In *History of the Expedition Under the Command of Captains Lewis and Clark* . . . Philadelphia, 1814.)

The Lewis and Clark expedition took place between 1804 and 1806, a decade before publication of the narrative and of this map. They explored the Louisiana Territory, ascending the Missouri, crossing over to the Columbia, where they descended to the Pacific. This is an epochal map which profoundly influenced not only the cartography of the West but the development of that region for many decades. Lewis and Clark were the pioneers in the Northwest, and everything they saw or heard about, they placed on their map! This remains the greatest cartographic monument to their most successful expedition which justified the great faith placed in them by the visionary Thomas Jefferson.

25) "A Map of the Internal Provinces of New Spain." Zebulon Pike.
(In *An Account of Expeditions to the Sources of the Mississippi, and Through the Western Parts of Louisiana* . . . Philadelphia, 1810.)

Before Lewis and Clark returned from their journey to the Pacific, Zebulon Pike, at 27 years of age and a lieutenant in the United States Army, had been sent westward by General James Wilkinson, Governor of the Louisiana Territory. He was to investigate the headwaters of the Arkansas and Red Rivers and possibly, Wilkinson imagined, his travels might take him near the settlements of New Mexico. Before this journey was over, Pike was captured by the Mexicans, held prisoner, and subsequently released.

This map is almost a direct plagiarism of Humboldt's great map of New Spain published in 1811 which we have already discussed. How did Pike acquire this map and include his derivative of it in the account of his expedition published in 1810? Humboldt tells us himself in another of his books, when mentioning his map of New Spain, that he had given a manuscript copy of it to Secretary of State

James Madison in Washington in 1804. This copy was no doubt officially issued to Pike before his departure.

26) "Map of the Country Situated Between the Meridian of Washington City and the Rocky Mountains...." Stephen H. Long. 1820.

Major Long explored the front range of the Rocky Mountains and discovered what was soon to become known as Long's Peak. This manuscript wall map borrowed much of its eastern portions from Tanner's more detailed maps.

Long was responsible for the term "Great American Desert" and was the father of this persistent myth regarding the Great Plains. His statement on "roving bands of Indians" caused much public apprehension and did little to encourage settlement of the area for a generation. Together with Lewis and Clark, and to a lesser extent Pike, these were the "mother maps" of the West until the period of Fremont two decades later.

Derivatives of this map were published in 1822 by Carey & Lea in their *A complete historical, chronological and geographical American atlas* and in 1823 by Edwin James in his *Account of an expedition from Pittsburgh to the Rocky Mountains....*

27) "Map of the United States of North America with Parts of the Adjacent Countries." David H. Burr, 1839.
(In Burr, David: *The American Atlas Exhibiting the Post Offices*.... London: J. Arrowsmith, 1839)

This is the first publication of the remarkable Jedediah Smith information which unraveled many of the secrets of the West. Smith traveled throughout the American West from 1826-30 and was the first explorer of the Great Basin. He was the first American to make his way into California from the east and back again overland. He crossed the Sierra Nevada and discovered the interior valley east of the Coastal Range. All of this is reflected on Burr's map, which has a number of specific references to Smith's routes.

Smith was killed at the early age of 33 by Comanches on the Santa Fe trail. If his own map had been published he would certainly have achieved an even greater stature in the exploration of the American West. His friend and former partner, Gen. Ashley, the celebrated St. Louis fur trader, was a member of the House of Representatives from 1831 to 1837 and apparently it was he who offered the opportunity to David Burr, Geographer to the House of Representatives, to examine a Jedediah Smith map. There is no doubt that Burr had a Smith map as his travels are developed in considerable detail and Smith's own legends and place names are carefully shown.

28) "Map of an Exploring Expedition to the Rocky Mountains in the Year 1842 and to Oregon & North California in the Years 1843-44 . . . ." Charles Preuss. (In Fremont, J.C.: *Report of the Exploring Expedition to the Rocky Mountains* . . . . Washington: Gales and Seaton, 1845. 28th Cong., 2nd Sess., Sen. Ex. Doc. 174, Serial 461.)

This map covers the area from the mouth of the Columbia, south to Los Angeles and is the most accurate map of the American West for its time. Fremont was the first to circle the entire Great Basin. His inscription in the vertical arc across the Basin reads: "The Great Basin: diameter 11 degrees of latitude by 10 degrees of longitude: elevation above the sea between 4 and 5,000 feet surrounded by lofty mountains; contents almost unknown, but believed to be filled with rivers and lakes which have no communication with the sea, deserts and oases which have never been explored, and savage tribes, which no traveller has seen or described."

Undoubtedly, this map made a lasting contribution to the cartography of the American West. It was on this map that the pioneer Oregon cartographer, George Gibbs, drew the Jedediah Smith data which Dale Morgan and Carl Wheat have described in fascinating detail in their monograph *Jedediah Smith and his maps of the American West* accompanied by excellent reproductions.

Charles Preuss actually drew this and many other important maps for John Charles Fremont. Preuss was the most important map-maker in the West during the 1840's. Born in Germany, he was usually an unhappy man during his adult life in America, in spite of what seems to have been an exciting career. "It is certainly terrible," wrote Preuss in his diary at Walla Walla with Fremont on October 27, 1843, "what a poor devil has to contend with in this country in order to make an honest living." He took his own life at 51, apparently never sharing Fremont's love of being on — or off — the trail.

29) "Map of Oregon and Upper California from the Surveys of John Charles Fremont and Other Authorities." Charles Preuss. (In Fremont, John Charles: *Geographical Memoir Upon Upper California, in Illustration of His Map of Oregon and California.* Washington: Wendell and Van Benthusen, 1848. 30th Cong., 1st Sess., Senate Misc. Doc. No. 148, Serial 511.)

While there are still a few aberrations, I feel it is quite amazing how we can see the comparative quickness in which the great empty spaces of the American West were being filled in in the final years of the first half of the nineteenth century.

Because of the circumstance, the Fremont-Preuss 1848 map is one of the first to indicate the location of the Gold Region in California. It is also the first map to use the term Golden Gate, at San Francisco. Despite its few defects, the map is an excellent delineation of the expedition of 1845-46.

SUMMARY

We have noted and discussed examples representing 400 years of mapping: from the fanciful visions of the sixteenth century to the scientific surveying of the nineteenth and so we pass from the decorative creatures in the oceans to an accurate description of a vast expanse of land known as the American West. Although there is no going back, we are fortunate still to have the examples of these explorers, cartographers and engravers to study and daydream of those early times.

SELECTED REFERENCES

Adams, James Truslow, ed. *Atlas of American History*. New York, 1943.

Bagrow, Leo. *History of Cartography*. Rev. and enlarged by R. A. Skelton. Cambridge, Mass.: Harvard, 1964.

Berkman, Brenda & Robert W. Karrow, Jr. *Index to Maps in the Catalogue of the Everett D. Graff Collection of Western Americana*. Chicago: Newberry Library, 1972.

Brown, Lloyd A. *The Story of Maps*. Boston: Little, Brown, 1949.

Brown, Lloyd A. comp. *The World Encompassed*. Baltimore: Trustees of the Walters Art Gallery, 1952.

Burrus, Ernest J. *Kino and the Cartography of Northwestern New Spain*. Tucson: Arizona Pioneers Historical Society, 1965.

Cowan, Robert Ernest & Robert Grannis Cowan. *A Bibliography of the History of California 1510-1930*. San Francisco: J. H. Nash, 1933. Four volumes.

Cumming, W.P., R.A. Skelton, & D.B. Quinn. *The Discovery of North America*. London: Elek, 1971.

Cumming, W.P., S.E. Hillier, & D.B. Quinn. *The Exploration of North America, 1630-1776*. New York: Putnam, 1974.

Diller, Aubrey. "A New Map of the Missouri River drawn in 1795," *Imago Mundi* XII, pp. 175-80. Stockholm, 1955.

Fite, Emerson D. & Archibald Freeman, eds. *A Book of Old Maps Delineating American History from the Earliest Days Down to the Close of the Revolutionary War*. Cambridge, Mass.: Harvard, 1926.

Friis, Herman R., ed. *The Pacific Basin*. New York: American Geographical Society, 1967.

Goetzmann, William H. *Army Exploration in the American West 1803-1863*. New Haven: Yale, 1959.

Goetzmann, William H. *Exploration and Empire*. New York: Knopf, 1966.

Harrisse, Henry. *The Discovery of North America*. London: H. Stevens, 1892.

Johnson, Adrian. *America Explored*. New York: Viking, 1974.

LeGear, Clara Egli, comp. *United States Atlases: A List of National, State, County, City and Regional Atlases in the Library of Congress.* Washington: U.S. Library of Congress, 1950-53. Two volumes.

Leighly, John. *California as an Island.* San Francisco: Book Club of California, 1972.

Lowery, Woodbury. *The Lowery Collection: A Descriptive List of Maps of the Spanish Possessions Within the Present Limits of the United States, 1502-1820.* Edited by Philip Lee Phillips. Washington: U.S. Government Printing Office, 1912.

Morgan, Dale L. *Jedediah Smith and his Maps of the American West*, San Francisco: California Historical Society, 1954. (California Historical Society. Special Publication no. 26.)

Paullin, Charles O. *Atlas of the Historical Geography of the United States.* Edited by John K. Wright. Washington: Carnegie Institution of Washington, 1932.

Phillips, Philip Lee. *A List of Maps of America in the Library of Congress.* Washington: U.S. Government Printing Office, 1901.

Reps, John W. *Cities of the American West.* Princeton: Princeton, 1979.

Schwartz, Seymour I. and Ralph E. Ehrenberg. *The mapping of America.* New York: Harry N. Abrams, 1980.

Shirley, Rodney W. *The mapping of the World: early printed World maps, 1472-1700.* London: Holland Press, 1983.

Skelton, R.A. *Explorers' Maps.* London: Routledge & Paul, 1958.

Storm, Colton, comp. *A Catalogue of the Everett D. Graff Collection of Western Americana.* Chicago: University of Chicago Press, 1968.

Tooley, R.V. "California as an Island," *Map Collectors' Circle,* Number Seven. London, 1963.

Tooley, R.V. "French Mapping of the Americas. The De l'Isle, Buache, Dezauche Succession (1700-1830)," *Map Collectors' Circle,* Number Thirty-Three. London, 1967.

Tooley, R.V., Charles Bricker & Gerald Roe Crone. *Landmarks of Mapmaking.* Amsterdam: Elsevier, 1968.

Tooley, R.V. *Maps and Map-Makers.* Sixth Edition. New York: Crown, 1978.

Vindel, Francisco. *Mapas de America en los Libros Espanoles.* Madrid, 1955.

Wagner, Henry R. *The Cartography of the Northwest Coast of America to the Year 1800.* Berkeley: University of California Press, 1937. Two volumes.

Wagner, Henry R. *The Spanish Southwest 1542-1794.* Albuquerque: The Quivira Society, 1937. Two volumes.

Wagner, Henry R. & Charles L. Camp. *The Plains and the Rockies. A Bibliography of Original Narratives of Travel and Adventure 1800-1865.* Third edition. Columbus, Ohio: Long's College Book Co., 1953.

Wheat, Carl I. *Mapping the American West 1540-1857. A Preliminary Study.* Worcester, Mass.: American Antiquarian Society, 1954.

Wheat, Carl I. *Mapping the Transmississippi West.* San Francisco: The Institute of Historical Cartography, 1957-58. 5 volumes in 6.

Wheat, Carl I. *The Maps of the California Gold Region.* San Francisco: Grabhorn Press, 1942.

Wheat, Carl I. "Twenty-five California Maps," *Essays for Henry R. Wagner.* San Francisco: Grabhorn Press, 1947.

Wilford, John Noble. *The Mapmakers.* New York: Knopf, 1981.

Wroth, Lawrence C. "The Early Cartography of the Pacific," *The Papers of the Bibliographical Society of America,* Volume Thirty-eight, Number Two. New York, 1944.

THE NOTORIOUS DOCTOR ROBINSON: A MEXICAN REVOLUTIONARY'S MAP OF NORTH AMERICA

Robert Sidney Martin
Louisiana State University
Baton Rouge

One *of the greatest diplomatic*

triumphs of Thomas Jefferson's presidency was his peaceful acquisition of the Louisiana Territory from France in 1803, an acquisition which more than doubled the size of the young republic. Yet, as Carl Wheat has pointed out, the Louisiana Purchase was a geographic pig-in-a-poke: the western boundary of the new territory was at best ill-defined and the lands it comprised as yet unknown.[1] Jefferson himself eventually came to the conclusion that Louisiana extended to the Rio Grande. This conclusion was based on the French claim to Texas resulting from the La Salle expedition, and confirmed by the subsequent French grant of much of that area to Antoine Crozat. Spain, on the other hand, held that her territory in North America extended to the Mississippi, and struggled to limit any extension of the United States west of that river.[2] The matter remained to be settled by direct negotiations between the United States and Spain. As these negotiations were repeatedly deferred by international conflicts in Europe, both sides set out to affirm their respective claims through explorations and settlement.

Against this backdrop unfolded some of the most inspiring, interesting, and intriguing episodes in the history of the American West. Participants in these actions bear names unforgettable to the student of American history, names like Lewis, Clark, Pike, Wilkinson, and Burr. One name which, although largely unknown, deserves to be added to this list is John Hamilton Robinson.

Little is known of Robinson's early life. He was born in Augusta County Virginia in 1782, and educated as a physician. Like many young men of his age, he saw adventure and opportunity in the vast new territories of the west. Upon completing his studies in Philadelphia in 1804 he moved to St. Louis, the gateway to the Louisiana Territory.[3] In 1805 he came to the attention of General James Wilkinson, who appointed him acting surgeon for the new establishment at Cantonment Belle Fontaine, just outside of St. Louis.[4] When Wilkinson dispatched Lieutenant Zebulon Montgomery Pike on his famous southwestern expedition in the summer of 1806, Robinson was added at the last moment as the expedition's surgeon and naturalist.[5]

Pike's expedition has been the subject of studies by a number of eminent historians. Yet it remains an object of intense interest because its interpretation must remain tentative, even speculative.[6] Robinson's role in the expedition has not gone unnoticed; indeed, he provides the focus for much of the speculation concerning the expedition's true purpose. He has been assigned various roles by historians: that of agent for Wilkinson in his clandestine dealings with the Spanish; that of Burr's representative in the expedition; and even that of *de facto* leader of the expedition. Pike's editor and biographer, Elliott Coues, flatly states that Robinson was a spy.[7] The evidence to support these contentions remains equivocal, but one interpretation fitting the facts is the simplest one: that he was an adventurous young man on the make, willing to grasp any opportunity to get ahead.

Whatever his motivation, there can be little doubt that Robinson played an active, even crucial, part in the expedition as it unfolded. He was constantly forging ahead of the little band to reconnoiter; he was frequently left in command in Pike's absence; his hunting skills often were the difference between starvation and survival. Pike came to rely increasingly on Robinson's presence, skill, judgement and determination as his party suffered increasing hardships.

In view of Robinson's subsequent career there are two main points to be made concerning his activities with Pike. First, during the course of the expedition, Robinson came into contact with a number of important individuals in New Spain. Among these were Nemesio Salcedo, the Captain General of the Internal Provinces; Juan Pedro Walker, an American-born engineer and surveyor in the Spanish service; Antonio Cordero, the Governor of Texas; and Simon Herrera, the military chief in Texas.

The most important of these was Salcedo, who distrusted Robinson from the start. One reason for his distrust was that, while in Chihuahua, Robinson made a clumsy attempt to defect. He wrote two letters to Salcedo, asking to be allowed to remain behind and become a Spanish subject. He outlined a grandiose scheme to explore the Northwest territory for Spain, and he claimed already to have the authorization of the American congress to plant a colony in Oregon. Salcedo refused his request and Robinson left Mexico with the rest of the party.[8]

The second important point is that Robinson returned to the United States with the general impression that the Spanish colonial government was feckless, corrupt, and tyrannical. He was already sympathetic to the revolutionary cause in Mexico.[9]

In his published journals, Pike speaks very highly of Robinson. He calls him "a young gentleman of science and enterprize," and notes that he had "blooming cheeks, fine complexion, and a genius speaking eye."[10] Relatively unlettered himself, Pike was impressed by Robinson's erudition:

> He has had the benefit of a liberal education, without having spent his time as too many of our young gentlemen do in colleges, viz. in skimming on the surfaces of sciences, without ever endeavoring to make themselves masters of the solid foundations, but Robinson studied and reasoned; with these qualifications he possessed a liberality of mind . . . [and] his soul could conceive great actions, and his hand was ready to achieve them; in short, it may truly be said that nothing was above his genius, nor anything so minute that he conceived it entirely unworthy of consideration. As a gentleman and companion in dangers, difficulties and hardships, I in particular, and the expedition generally, owe much to his exertions.[11]

To his chief, General Wilkinson Pike acknowledged that Robinson was "my companion in dangers and hardships, counselor in difficulties," and he added that for Robinson's "chymical [sic], botanical, and mineralogical knowledge the expedition was greatly indebted . . . ."[12]

Pike may have been impressed by Robinson's scientific accomplishment, but the natural history results of the expedition were negligible. Robinson may indeed have made numerous observations of the flora and fauna, as Pike claimed, but they remain unknown today. As for cartography, the only contribution Robinson made was to copy some of Pike's sketches and notes, thus preserving them when the Lieutenant's papers were confiscated.[13]

His experiences with Pike seem to have whetted rather than sated Robinson's thirst for adventure. For the next several years he appears a man in search of himself. He had married an attractive young woman of French birth before setting out from St. Louis with Pike, and upon his return he apparently tried to settle down with his family. He tried his hand at a number of different enterprises, but made his mark in none of them. He continued as an acting surgeon at Belle Fontaine for a time.[14] Using the good offices of his friend Pike he attempted to wangle a permanent appointment in the Army, and in December, 1808, one John Hamilton Robinson was in fact appointed to the post of ensign in the Second Regiment of Infantry. But Robinson never served in that capacity; perhaps the rank was not exalted enough.[15]

At the same time, Robinson was also attempting to secure an appointment in the Indian Department. In 1809 this campaign bore fruit and he was posted as a sub-agent to Fort Osage, where he became associated with George C. Sibley, the Indian Agent at the post. Together they were involved in a contretemps with the commanding officer that led, ultimately, to Robinson's banishment from the post and the surrounding territory.[16] From there Robinson went to Kaskaskia, Illinois, where he entered into business with his brother-in-law. Soon he was appointed aid de camp to William Rector, Brigadier General of the Illinois Militia.[17] But Robinson was not satisfied with the humdrum life of a frontier merchant, even with the added bonus of a part-time military appointment. Prompted perhaps by the momentous events of the Mexican Revolution then under way, as well as the impending war with Britain, in 1812 Robinson again importuned his friend Pike, by now an extremely influential colonel, to aid him in securing some kind of important work.

Pike arranged a meeting for Robinson with Secretary of State James Monroe, where they discussed the subject of the Internal Provinces. Facing a difficult conflict with Great Britain, Monroe was anxious to secure the nation's southwestern flank, and he was concerned that the filibusterers and revolutionaries then active on the

Texas border might precipitate war with Spain. He asked Robinson to submit his ideas on that topic through Pike. The doctor drafted a letter for Pike's signature outlining the situation in New Spain, and strongly recommending himself as an envoy to Nemesio Salcedo.[18] On July 1, 1812, Monroe appointed Robinson to that post and instructed him to present the views of the United States concerning the activities of the filibusterers in the neutral ground. He was to suggest a need for mutual action to eradicate these banditti, and take every opportunity to promote commerce and friendly relations between the United States and New Spain. He was also to tell Salcedo that the United States viewed the boundary question as one for settlement through amicable negotiations.[19]

Robinson proceeded to Natchitoches, where he arrived in October 1812. En route across Texas, on the upper Trinity River, he encountered the main group of insurgents under Bernardo Gutiérrez de Lara and Augustus Magee. The rebels were suspicious that Robinson had come to take possession of Texas for the United States, and they held him captive until he vowed in writing not to reveal to Spanish authorities the location and strength of the rebel forces. Further along the route he met Miguel de Salcedo, the Captain General's nephew, and renewed his acquaintance with Simon de Herrera and Antonio Cordero. Upon reaching Chihuahua he presented his credentials to Salcedo, but succeeded only in infuriating the Captain-General, who suspected that Robinson came actually to foment revolution, not to quell it. There was good reason for Salcedo to suspect Robinson's motives; not only had his behavior on his previous visit to Chihuahua been suspicious, but while now in the city Robinson openly met with representatives of various Revolutionary juntas. Claiming Robinson's authority insufficient, Salcedo refused to deal with him. Robinson returned via the same route he had come, and took every opportunity to meddle in local affairs and to generally make a nuisance of himself.[20]

Robinson returned to the United States thoroughly committed to the Mexican Revolution. Along the way back to Washington he met José Alvarez de Toledo, the newest leader of the revolutionary juntas in Louisiana. The two men quickly became intimate friends, and set about developing plans for an invasion of the Internal Provinces.[21]

When Robinson returned to Washington to report to Monroe he used the opportunity to try to secure support for their revolutionary activities.

His report to Monroe runs to forty pages of grandiloquent prose, giving a detailed account of his activities while in Mexico and a verbatim transcription of his conversations with Salcedo. To this he appended his personal views on the "Present State of the Mexican Revolution," which is a glowing endorsement of the cause of Mexican Independence. His final paragraph will suffice to illustrate his orotund style and his magnificent revolutionary vision:

> This revolution [in Mexico] is the more important since on its destiny depends the Liberty and Independence of all South America. . . . I have not, in my mind, limited the liberties of Americans by the Isthmus of Darien or even the Amazonian Mountains, but have viewed the whole space comprehended between Louisiana and Cape Horn, divided into many free and independent governments, enjoying the blessings of liberty and peace, and holding the oldest Republic of America by the hand of friendship, bid defiance to the storms and thunder of European tyrants.[22]

The effects of this manifesto on Monroe may well be imagined: he had dispatched Robinson to Mexico in order to smooth relations with the authorities in New Spain; the flamboyant physician had instead stirred up a hornet's nest, and was now urging the United States actively to support the revolutionaries in Mexico. Robinson was politely thanked for his views, and calmly advised that his services were no longer needed by the State Department.[23]

The good Doctor was apparently oblivious to the effect he had in Washington, however. He convened a military junta in Philadelphia and issued a manifesto, calling upon the men of America to enlist in the Mexican revolution. This proclamation appeared as a broadside, and was reprinted in a number of newspapers in the West. In it he extolled the virtues of the revolution in language similar to that in his report. He invoked the name of his friend Pike, by now a casualty of the War of 1812 and a national hero, claiming that "the late gallant and brave General Pike would have been with us, had he lived."

Robinson had the temerity to mail these broadsides to a number of important citizens throughout the Western states, inscribed "Rendezvous at Nacogdoches, February 25."[24] This led to a number of inquiries from the recipients of the manifesto, and Monroe was forced to disavow Robinson and issue writs for Robinson's arrest. There is some doubt that Monroe actually wanted to apprehend Robinson, since he easily evaded arrest, even when in the company of a federal marshall.[25]

From Philadelphia Robinson proceeded to Natchez, where he convened another junta and was commissioned to lead its forces in an invasion of the internal provinces. He arrived in Texas in February, 1814, where he came into direct conflict with Toledo over who was to command the revolutionary force. He was arrested by Toledo's men, tried on a charge of conspiracy, and acquitted. The two men patched over their differences, but they never trusted each other again. Throughout the rest of the year they competed for recruits and maintained separate camps, neither making any attempt to actually mount an invasion.[26] In the fall of 1814, with the impending British invasion of Louisiana, Robinson offered the services of his volunteers to help defend New Orleans. This offer was politely declined by William Claiborne, the governor of Louisiana, who held a warrant for Robinson's arrest. The intrepid Doctor nevertheless went alone to New Orleans, where he was commissioned a surgeon in the militia and placed in charge of a hospital near New Orleans.[27]

The war marked only a temporary hiatus, not an end, to Robinson's revolutionary activities. Shortly after the Battle of New Orleans, Robinson was again drawn into a junta, and found himself on his way to Mexico as its representative to the Mexican revolutionaries. He sailed in the summer of 1815 for Vera Cruz, and from near there dispatched another of his grandiloquent epistles, this addressed to President Madison himself, enclosing a copy of the new Mexican constitution. For the next eighteen months he was directly involved in the affairs of the revolution, serving with the likes of Mier y Terán and Guadalupe Victoria. For his service he received a commission as Brigadier General in the Revolutionary Army, and a stipend purportedly amounting to $20,000.[28]

Robinson's activities did not go unnoticed by the officials of Spain. Don Luis de Onís, the Spanish Charge d'affair in Washington, wrote repeatedly to Monroe, expressing outrage that the United States did nothing to stop the activities of the revolutionary agents operating on its soil, and asking that they be punished as violators of international law and the neutrality of the United States. He singled out Gutiérrez, Toledo, and Robinson as the most egregious offenders. "As regards Doctor Robinson," Onis wrote,

> it is notorious that he has been one of the most infuriated enemies of Spain, and the one who has, with the greatest eagerness, promoted the rebellion of the provinces of His Majesty. It was he who introduced himself into the internal provinces to seduce their inhabitants; it was he who sowed the seed of insurrection; it was he who procured intelligence in San Antonio . . . ; and it was he who published, in these United States, proclamations, signed with his hand, inviting adventurers from all parts to form an army . . . , and, in one word, declaring war himself . . . against the Spanish nation . . . .[29]

These charges were published, and Robinson was not the kind of man to let them go unanswered. He responded in kind, and made certain that his reply was widely distributed. In it he rebuked Onis personally: "Your language, sir, is extremly indecorous, [and] I shall not descend from the dignity of an American and an officer of the Mexican republic to answer the illiberal and scurrilous observations of the minister of Ferdinand VII." The king himself Robinson mocked for being "too much occupied with the organization of the officers of the bedchamber and toilette apartments of his young spouse to attend at this moment to the trifling considerations of Spanish relations with the United States, or the insurrection of a hundred provinces in America."[30]

Robinson returned to the United States in the spring of 1817, his health broken from fatigue and exposure. He settled in Natchez with his family to recuperate, and from there he continued to shower Monroe and Madison with reports on the revolution and requests for support.[31] By early 1818 his health was sufficiently restored for him to begin work on his next great project, a map of the territories through which he had journeyed and fought during the previous decade. In March of that year he issued a Prospectus for such a map, and solicited subscriptions at fifteen dollars each.[32]

Robinson claimed that the map he planned to produce would be based on the knowledge he had acquired during his service in Mexico, and would utilize a number of official Spanish sources. The resulting map, he claimed, would be "the most perfect which has appeared before the public." It "will contain the latest and best information from the discoveries and possessions of the American, Spanish, Russian, British and French travellers and navigators, and representing the claims of their respective governments in the North western coast of America."[33]

Robinson devoted the ensuing months to compiling the map, apparently completing the task during the summer of 1818. He spent the remainder of that year travelling between Natchez, Washington, and Philadelphia, arranging to have it published. By January, 1819, he had collected some 400 subscriptions, sufficient to underwrite the engraving and printing. The map was issued for distribution sometime during February, 1819.[34]

The final product of these exertions is a large wall map,[35] printed on six sheets which, when joined, measure approximately five and a half feet square. The map covers the greater portion of Western North America. It features a large, ornate title block, with an allegorical representation of Liberty showering her blessings on the United States and the Mexican Republic, while a humble soldier — perhaps Robinson himself — bows in grateful salute. The map was engraved by Hugh Anderson, a well-known if minor Philadelphia engraver of maps and portraits, and it was printed by John L. Narstin of Philadelphia, about whom little is known.[36] The map is dedicated to seven men — all minor political figures in Natchez and the nearby Mississippi Territory — in gratitude for their support in securing its publication.

On the face of the map appears the statement "The information on which the author feels himself justified in the publication of this map is from his own knowledge of the country in his several voyages thither and also the several manuscript maps which are now in his possession, drawn by order of the Captain General of the Internal

Provinces and the Viceroy of Mexico." As to what manuscript maps Robinson used we are given but one clue: along the Pacific coast we find the prominent legend: "This portion of the coast was laid down from a map made by don Juan Pedro Walker by order of the Captain General in 1810." Walker was an American-born surveyor and engineer, a veteran of Andrew Ellicott's survey of the northern boundary of Florida, who had found a career in the Spanish service. During most of his career he was attached to the household of the Captain General of the Internal Provinces, Nemesio Salcedo, and it was in that capacity that he came into contact with Robinson, first when the Doctor was in Chihuahua with Pike in 1807, and again during Robinson's embassy to Salcedo in 1812-13.[37] Robinson may have also met with Walker on his subsequent visits to Mexico, but there is no evidence that he did so. It is possible that Walker may have supplied Robinson with the other maps that he claimed to possess, since Walker undoubtedly had access to such material. In any event, the coast line thus labeled does appear similar to that on a Walker map of western North America that has been assigned the date 1817.[38]

The map also features a number of statistical tables, two devoted to the size and location of various Indian tribes shown on the map, and others listing mountain peaks, mines, and the like. One of these tables gives the latitude and longitude of a number of key points "from actual observation," although whose observation is not specified. It is clear that Robinson did not make astronomical observations at all of these points, if indeed he made any. A cursory examination indicates they conform rather closely to those published in Humboldt's *Political Essay*.[39]

Wheat suggests that Robinson drew not only from Walker, but also from Lewis and Clark, Pike, and some Spanish sources that must have included Font, Miera, and perhaps Lafora.[40] There is no reason to question this assertion, for the map is replete with details derived from these authorities. The route of Lewis and Clark across the northwest is shown, along with numerous details of their journey. Likewise the routes of Font, Garcés, and Domínguez in the southwest are laid down. There are many notations that reflect the

activities of Pike's 1805 expedition in search of the source of the
Mississippi.

The most interesting of these details, however, are those relating to
the 1806-07 expedition of which Robinson was a member. The
route of the expedition is carefully traced, and along it are numerous
comments on the topography, the flora and fauna, and the activities
of the explorers en route. Along the Arkansas, for example, we find
notes indicating the point at which Lieutenant James Wilkinson was
detached with a small party to descend and reconnoiter that river by
boat. Further along is the note that "Here the Mountains are First
seen," and yet further a "Silver Mine discovered by Jas. Pursley in
1803" is marked. Other notes indicate the locations where Pike
constructed his stockade, and where he was captured by the Spanish
troops. The route of the Spanish company of Dragoons under
Lieutenant Facundo Melgares which had been dispatched to intercept
Pike and turn him back is also shown.

Natural features along this route are also noted. Among these are
notes along the Arkansas stating that "this stretch of the River is
covered with Islands;" "Particles of Salt seen here on the Surface by
Captain Pike in 1806;" and, near El Paso, "Excellent Wines made
Here." There are many references to springs, herds of wild horses,
and "Immense Herds of Buffalo." Perhaps the most notable such
indication of natural features is the prominent mountain near the
headwaters of the Arkansas, labeled "Pike's Mountain, 10,581 Feet
from its Base." It is fitting that here, for the first time, the peak is
named for the intrepid explorer with whom Robinson had attempted
its ascent unsuccessfully. Pike himself had merely labeled it "Highest
Peak" on his own map.[41]

Aside from these interesting historical notes, however, the map is
a disappointment; more than one historian has commented on its
deficiencies.[42] It is full of geographical apocrypha. Prominently
displayed, for example, is the "Mountain of the Prairie," in present
Minnesota. The depiction of the Great Basin is reminiscent of that
on maps a century earlier. Lake Timpanogos is shown, and the
Timpanogos, Buenaventura and San Felipe rivers all flow to the
Pacific unhindered by any hint of the Sierra Nevada. In Texas, the
depiction of the Sabine and the Red River are quite good, but

otherwise it is even cruder than Pike's map. The headwaters of the Brazos are, for example, confused with the Colorado, and those of the Red with the Canadian. Considering Robinson's experience, and his claimed access to information, the map is a failure as a contribution to the cartography of the region. One experienced frontiersman in Santa Fe, Josiah Sibley, confided to his journal in 1826 that the map was "not worth taking home. If offered for sale it would not bring a dollar here."[43]

To evaluate Robinson's map on these grounds, to label it — as Wheat does — a mere "curiosity," is to fail to place it in the proper perspective. Robinson's map is, in fact, of enormous importance, not geographically or cartographically, but politically. In order to appreciate its significance, it is important to place it in the context of Robinson's career as a dedicated revolutionary and meddlesome would-be diplomat.

Robinson's motives for undertaking the compilation and publication of this map are complex. To begin with, there is some indication that he expected to generate a substantial income from it. It is also clear that he thought himself to be in possession of much genuinely new geographic information, of interest to the general public.[44] Beyond this, however, it seems probable that his real motivation was political: the map was simply another way to carry on the revolutionary struggle to which he had devoted his life and health.

In early 1819, as Robinson was making the final preparation for the publication of his map, the protracted and long-delayed negotiations between the United States and Spain over the boundary between their respective territories in North America were drawing finally to a close. At the beginning, the United States had held to the position that its territory extended to the Rio Grande, with an indefinite line extending north and west from the headwaters of that river as far as the Pacific. Spain, on the other hand, contended that Texas was an integral part of the Internal Provinces, and that its eastern boundary was the Mississippi. In late January, 1819, the two sides, represented by Secretary of State John Quincy Adams and Minister Plenipotentiary Luis de Onís, were nearing a compromise as the negotiations progressed in Washington. By early February these representatives had reached a rough agreement that the line was to follow the Sabine River from its mouth to the 32nd parallel, thence due north to the Red River, up that stream to the 100th meridian, thence north to the Arkansas River, which it would follow to its source. It was from the source of the Arkansas that the remaining controversy lay. The Spanish suggested a line north from that point to the 43rd parallel, then due west to the Pacific; the United States held out for a line due west from the Arkansas to the Pacific, to run no farther north than the 40th parallel. Numerous proposals and counter-proposals were exchanged between February 1 and February 15, aimed at reaching a compromise on this point. Final agreement was reached on February 22, when the two ministers signed a treaty establishing the boundary described, with the line running north from the source of the Arkansas to the 42nd parallel. This treaty, now known as the Adams-Onís Treaty, was ratified by the United States Senate on February 24.[45]

Throughout these negotiations the western boundary of the Louisiana Purchase remained a matter of great public interest in the United States. Expansionists vociferously demanded a line as far west as could be maintained, and many upheld the right of the United States to the Rio Grande.[46] Given Robinson's background of revolutionary activity, his fierce antipathy for the Spanish government in North America, and his intense commitment to securing the blessings of liberty for the citizens of New Spain, there can be no doubt that he himself took great interest in the course of these negotiations. That interest is revealed clearly in his map.

On the map Robinson prominently displayed the territory over which the controversy raged. To the east, just west of the Mississippi River, runs a line highlighted in blue and labeled "Eastern Limits of the Spanish Claim." To the west, highlighted in red, runs a line along the 40th parallel to the headwaters of the Rio Grande, and then along that river south to the Gulf of Mexico. This line is labeled, below Santa Fe, as the "Western Limits of the United States." The vast territory between these two lines, the fate of which hung on the outcome of the negotiations, was thus laid before the

view of all. Lest anyone fail to notice it, Robinson included an entry
in the table of "References," which notes:

> The red line Commencing at the 40° of Latitude Thence East
> to the Rio del Norte, Thence down that River to the Gulf,
> constitutes the Western Limits of Louisiana. The Blue Line near
> Natchitoches represent [sic], the Eastern limits of the claim of his
> Catholic Majesty.

Here displayed at a glance was the vast territory at stake in the
negotiations even then under way in Washington. Here graphically
portrayed was the rich patrimony that the United States government
was preparing to abandon to the Spanish. That this was an inten-
tional contrast drawn by Robinson can scarcely be doubted. It seems
most probable that this alone constituted an important rationale for
Robinson's interest in seeing his map published.

In addition to this internal evidence on the map itself, there are
other indications of Robinson's interest in this issue. In July, 1818,
shortly after he completed the draft of the map, Robinson showed it
to a number of individuals, probably as a part of his effort to secure
subscriptions. It soon attracted public notice in the newspapers. A
brief item in *Niles Weekly Register,* captioned "Small Difference!,"
notes the existence of the map, and the only feature mentioned is the
depiction of the boundary claims:

> The boundary lines of the territorial claims of the United States
> and Spain are marked on this map. The tract of countries lying
> between the extreme extent of country so claimed, is estimated to
> contain *one thousand and twenty-four million, 982 thousand
> acres!* [47]

Robinson also showed off his map closer to home, and a lengthy
article discussing it appeared in the *Mississippi Republican* in August, 1818. The article, entitled "Boundaries of Louisiana," is quick
to note that "The first object which attracts the eye of an American,
in examining this map, is the astonishing difference in the claims to
territory between the United States and Spain. . . ." After describing
these claims, the article points out that, if the Spanish claim were
allowed to stand, the resulting Louisiana Territory would extend less
than 100 miles west of the Mississippi River. The newspaper went

on to castigate the officials of the United States for their pusillanimity
in defending the rights of the republic to Louisiana:

> We cannot forbear expressing surprise, when we see the pro-
> posed western limits of Louisiana, . . . by which the president
> offers to relinquish two-thirds of the territory claimed under the
> purchase of Louisiana, and our mind has been the more particu-
> larly impressed with this extraordinary proposition, when we
> recollect that under every administration since that purchase, it has
> maintained that our claim extended to the Rio del Norte. Now if
> that claim be found in justice, as we have a right to believe it was,
> the American people will cheerfully defend it.
>
> The territory proposed to be relinquished by the president,
> contains about 697,216,000 acres of land . . . This territory is
> greater than that of the United States prior to the acquisition of
> Louisiana.[48]

There can be little doubt that these sentiments represented Robinson's own, nor that he was anything but pleased at the conclusions
drawn so readily from his handiwork.

The manuscript draft of the map which elicited these reactions is
not extant, but the surviving exemplars of the published version
provide even more evidence of the importance that Robinson placed
on the boundary issue. There are at least three variant states of the
map, and the points of variance all relate to the depiction of the
Adams-Onís boundary.[49]

It should be remembered that the final preparations for the
publication of the map were taking place in Philadelphia in February
1819, even as the Adams-Onís negotiations were drawing to a close
in near-by Washington. The earliest state of the printed map
indicates that it may have issued before final agreement was reached
on the boundary: in addition to the lines showing the rival claims of
the two powers, already described, this copy features a legend along
the 40th parallel, the Arkansas, and the Red River reading "Western
Limit of the United States" (along the fortieth parallel), and
"According to the Late Treaty" (along the Arkansas and the Red).
This boundary is highlighted in red, and corresponds to one then
under consideration in Washington.[50]

The second state of the map is a hybrid; the old part of the legend along the 40th parallel has been incompletely effaced from the plate — thus indicating that it is a subsequent state — and a new legend has been engraved along the forty-second parallel. This legend reads "Western Limits of the United States;" the remainder of the legend, "According to the Late Treaty," still appears as it was in the first state. Interestingly, both the 40th and the 42nd parallel lines were highlighted by the colorist.[51]

The third state of the map has the same boundaries as the second state, but it features a significant addition: along the Rio Grande, just south of Santa Fe, the single word "Former" has been added to the legend there, so that it now reads "Former Western Limits of the United States."[52]

From this internal evidence it is clear that Robinson was attempting to keep pace with events as they unfolded even while his map was being printed. The alteration of the legend along the Rio Grande, rather than its erasure, implies that he was attempting to keep what he viewed as the rightful claim of the United States to much of New Spain in the public view.

There was, in fact, an upswelling of popular sentiment, particularly in the West, against the terms of the Adams-Onís Treaty. The filibusters and invasions to liberate Texas and northern New Spain did not stop. Indeed, it is interesting to speculate on what connection there might have been between Robinson and the expedition of James Long which was mounted in early 1819. Like Robinson, Long was a resident of Natchez, and his wife was the niece of Robinson's old patron, General James Wilkinson. Any connection between the two must remain speculative, however, in the absence of any direct evidence.[53]

Robinson, weakened by his exploits, succumbed to yellow fever in September, 1819, scarcely six months after the publication of his

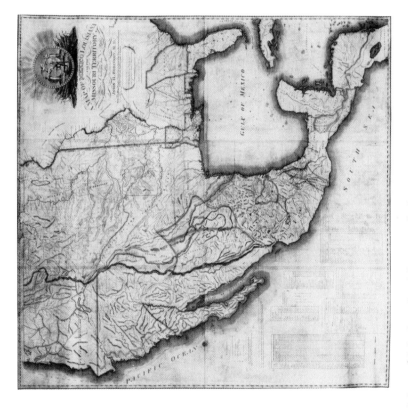

Fig. 7. "A Map of Mexico, Louisiana, and the Missouri Territory." John Hamilton Robinson.

map. He died at the age of 37. An obituary published by a sympathetic editor serves as a fitting epitaph:

> Doctor Robinson embarked at an early period with the lamented Gen. Pike, to explore the country west and south of the Arkansas, and there witnessing the degraded state in which his fellow creatures were doomed to drag out a wretched existence in so fair a portion of the globe; alive to every patriotic and philanthropic feeling, he pledged himself to the respectable citizens of Santa Fe and Chewawa, to use every means to arouse his fellow citizens of the United States to vindicate their rights, to aid in their emancipation from the most abject tyranny, and raise them to a political level with their brethren of the north. His pursuits have been constantly directed to this grand object and would have succeeded, had he met with men equal with himself in wisdom to plan, and courage to execute.[54]

Robinson's map, rather than a mere cartographic curiosity, must be seen as another of his "pursuits . . . directed to this grand object." His purposes were at least as much political as geographic. As a result the map must be interpreted within the context of Robinson's entire career and the broader historical framework in which it took place. Viewed in this light, it serves as an excellent example of the way in which maps have sometimes served other than cartographic ends.

NOTES

1. Carl I. Wheat, *Mapping the Transmississippi West*, 5 vols. (San Francisco: Institute of Historical Cartography, 1957-63), 2: *From Lewis and Clark to Fremont, 1804-1845*, (1958), pp. 1-2.

2. Thomas Maitland Marshall, *A History of the Western Boundary of the Louisiana Purchase, 1819-1841* (Berkeley: University of California Press, 1914), pp. 8-16; Thomas Jefferson, "The Limits and Bounds of Louisiana," in *Documents Relating to the Purchase and Exploration of Louisiana* (Boston: Houghton & Mifflin, 1904).

3. Harold A. Bierck, Jr., "Dr. John Hamilton Robinson," *Louisiana Historical Quarterly* 25(1942): 644; "John Hamilton

Robinson," (typed sketch), John Hamilton Robinson Papers, Missouri Historical Society; Louis Houck, *A History of Missouri*, 3 vols. (Chicago: R. R. Donnelley & Sons, 1908), 3:80; A. J. Morrison, "Dr. John Hamilton Robinson, 1732-1819 [sic]," *Tyler's Quarterly Historical Review and Genealogical Magazine 3* (January 1922): 154; Frederic L. Billon, *Annals of St. Louis in its Territorial Days from 1804 to 1821, Being a Continuation of the Author's previous work, The Annals of the French and Spanish Period* (St. Louis: privately printed, 1888).

4. Wilkinson to the Secretary of War, August 10, 1805, Clarence Edwin Carter, ed., *The Territorial Papers of the United States*, 28 vols., (Washington: Government Printing Office, 1934-   ), 13: *The Territory of Louisiana-Missouri, 1803-06*, pp. 181-82.

5. Wilkinson to Pike, July 12, 1806, published in Donald Jackson, ed., *The Journals of Zebulon Montgomery Pike, with Letters and Related Documents*, 2 vols. (Norman: University of Oklahoma Press, 1966), 1:288-89.

6. The most recent and comprehensive version of Pike's expedition is that of Jackson, cited above; in addition see: Elliott Coues, *The Expeditions of Zebulon Montgomery Pike, To the Headwaters of the Mississippi River, Through the Louisiana Territory, and in New Spain, During the Years 1805-6-7*, 3 vols. (New York: Francis P. Harper, 1895); Thomas Perkins Abernathy, *The Burr Conspiracy* (New York: Oxford University Press, 1954), Chapter 9, "Pike's Peek," pp. 119-37; Isaac Joslin Cox, "Opening the Santa Fe Trail," *Missouri Historical Review* 25 (October 1930), 30-66; W. Eugene Hollon, *The Lost Pathfinder: Zebulon Montgomery Pike* (Norman: University of Oklahoma Press, 1949).

7. Cox, "Opening the Santa Fe Trail," pp. 54-55; Abernathy, *Burr Conspiracy*, pp. 120-21; Henry Raup Wagner, *The Plains and the Rockies: A Bibliography of Original Narratives of Travel and Adventure, 1800-1865*, 3rd ed., revised by Charles L. Camp (Columbus, Ohio: Long's College Book Co., 1953), p. 25; Coues, *Expeditions*, 1:499.

8. Jackson, *Journals of Pike*, 2: 192-93, 204-06.

9. ibid., p. 245.

10. ibid., pp. xxiv, 402.

11. ibid., pp. 377-78.

12. ibid., p. 243.

13. ibid., p. 240. The historian of scientific exploration of the American West states that, had a genuine naturalist been appointed to Pike's expedition, rather than Robinson, "the natural history results of Pike's journeys might have been worth recording." Susan Delano McKelvey, *Botanical Exploration of the Transmississippi West, 1790-1850* (Jamaica Plain, Mass.: Arnold Arboretum of Harvard University, 1956), pp. 225-26.

14. Helene Foure Selter, *L'Odyssee Americaine d'une Familie Francaise, Le Docteur Antoine Saugrain: Etude suivie de Manuscrits Inedits et de la Correspondence de Sophie Michau Robinson* (Baltimore: Johns Hopkins University Press, 1936), p. 73-74; *American State Papers. Miscellaneous.* 2 vols. (Washington, 1834), 1:577-78.

15. Pike to Henry Dearborn, December 7, 1807, in Jackson, *Journals of Pike,* pp. 282-83; ibid., p. 381; William H. Powell, *List of Officers of the Army of the United States from 1779 to 1900* (New York, 1900), p. 47.

16. Robinson to Frederick Bates, September 29, 1808, Bates Collection, Missouri Historical Society; William Clark to the Secretary of War, April 29, 1809, *Territorial Papers,* 14: *The Territory of Louisiana-Missouri, 1806-1814,* p. 264; Kate L. Gregg, "The History of Fort Osage," *Missouri Historical Review* 34(1940): 446-55; E. B. Clemson to William Custis, July 20, 1810 (with enclosures), Records of the Office of the Secretary of War: Letters Received, C-195(5), National Archives; Clemson to Secretary of War, August 16, 1810, ibid., C-207(5); Secretary of War of William Clark, August 7, 1809, *Territorial Papers,* 14:289; ibid., p. 400, note 78.

17. Robinson to John Michau, Jr., (promissory note), April 27, 1811, Saugrain Papers, Missouri Historical Society; "Executive Register," June 4, 1811, *Territorial Papers,* 17: *The Territory of Illinois, 1814-1818,* p. 636; Foure Selter, *L'Odyssee,* p. 74.

18. Pike to James Monroe, June 19, 1812, Ford Collection, New York Public Library; reproduced in Jackson, *Journals of Pike,* 2:379-81. This letter is entirely in the hand of Robinson, except for the complementary close and signature, which are by Pike.

19. Monroe to Robinson, July 1, 1812, Department of State, *Correspondence Relating to the Filibustering Expedition against the Spanish Government of Mexico, 1811-1816* (Washington: National Archives Microfilm Publication, T286.)

20. Robinson to Monroe, July 20, 1813, *Filibustering Expedition.* See also: Bierck, "John Hamilton Robinson," pp. 651-56; Isaac Joslin Cox, "Monroe and the Early Mexican Revolutionary Agents," American Historical Association *Annual Report,* 1911, 1:208-215; Carlos Castañeda, *Our Catholic Heritage, 1519-1936,* 6 vols. (Austin: Von Boeckmann-Jones, 1950), 5: *Transition Period: The Fight For Freedom,* pp. 90-93; Mattie Austin Hatcher, *The Opening of Texas to Foreign Settlement, 1801-1821,* University of Texas *Bulletin* no. 2714 (Austin: The University of Texas, 1927), pp. 221-22; Julia Kathryn Garrett, *Green Flag over Texas: A Story of the Last Years of Spain in Texas* (New York: Cordova Press, 1939), pp. 163-69, 178; Harris Gaylord Warren, *The Sword Was Their Passport: A History of American Filibustering in the Mexican Revolution* (Baton Rouge: Louisiana State University Press, 1943), pp. 37-42.

21. Robinson to Monroe, July 20, 1813; Garrett, *Green Flag, p. 189.*

22. Robinson to Monroe, July 20, 1813.

23. Monroe to Robinson, June 25, 1813, *Filibustering Expedition.*

24. "Fellow Citizen," September 15, 1813, (printed broadsheet, with ms. note at top: "Confidential, Philadelphia, Sept. 15th 1813," and at bottom: "Health and Fraternity, John H. Robinson, Rendevous, at Nacogdoches 25th. Nov. 1813," both in Robinson's handwriting), Records of the Office of the Secretary of War, Letters Received, R-125(7), National Archives; another copy, similarly signed, dated September 18, 1813, Saugrain Papers, Missouri Historical Society, published in

Jackson, *Journals of Pike,* 2:382-87; a ms. transcript of a similar document appears in *Filibustering Expedition;* (Natchez) *Mississippi Republican,* February 23, 1814; Bierck, "John Hamilton Robinson," pp. 657-59; Warren, *Sword,* p. 76.

25. William Rector to the Secretary of War, October 28, 1813, *Territorial Papers,* 16: *The Territory of Illinois, 1809-1814,* p. 373; Secretary of State to Governor Edwards, January 21, 1814, ibid., 14:394; Secretary of State to William Mears, January 21, 1814, ibid., 14: 395; Secretary of State to Governor Clark, January 21, 1814, ibid., 14:733; James Monroe to Alexander James Dallas, December 15, 1813, Department of State, *Domestic Letters* (Washington: National Archives Microfilm Publication, M40), roll 14; Dallas to Monroe, January 10, 1814, Department of State, *Miscellaneous Letters of the Department of State, 1789-1906* (Washington: National Archives Microfilm Publication, M179), roll 29; Monroe to Dallas, January 15, 1814, *Domestic Letters;* Monroe to Robinson, February 14, 1814, ibid.; Monroe to the Governors of Louisiana and the Misisipi [sic] territory, February 14, 1814, ibid.; Dallas to Monroe, May 17, 1814, *Miscellaneous Letters;* Robinson to Antoine Saugrain, April 18, 1817, in Foure Selter, *L'Odyssee,* p. 78.

26. Bierck, "John Hamilton Robinson," pp. 661-62; Warren, *Sword,* pp. 82-83, 89-95; Everett S. Brown, ed., "Letters from Louisiana, 1813-14," *Mississippi Valley Historical Review* 11(1925): 572-79; Jose Alvarez de Toledo to William Shaler, May 30, 1814, with enclosures, *Filibustering Expedition.*

27. Bierck, "John Hamilton Robinson," pp. 662-63; Foure Selter, *L'Odyssee,* pp. 74-78; William Claiborne to John Perkins, October 21, 1814, in Dunbar Rowland, ed., *Official Letter Books of W. C. C. Claiborne,* 6 vols., (Jackson, Mississippi: Department of Archives and History, 1917), 6:283-84; Marion John Bennett Pierson, *Louisiana Soldiers in the War of 1812* (Baton Rouge: Louisiana Genealogical and Historical Society, 1963), p. 102.

28. Bierck, "John Hamilton Robinson," pp. 663-66; *Niles Weekly Register,* 9(December 23, 1815): 229; William Davis Robinson, *Memoirs of the Mexican Revolution: Including a Narrative of the Expedition of General Xavier Mina* ... (Philadelphia: privately printed, 1820), pp. vi-x; Foure Selter, *L'Odyssee,* pp. 76-80; Lucas Alaman, *Historia de Mejico, desde los Primeros Movimentos que Preparon su Independencia en el ano de 1808 hasta la Epoca Presente,* 5 vols., (Mexico, 1849-52), 4(1851): 393-94, 439, Apendice Num. 8, p. 14; Robinson to Madison, July 3, 1815, *Miscellaneous Letters.*

29. *American State Papers. Foreign Relations.* 4 vols. (Washington, 1838-52), 4:426-29.

30. Robinson to Onis, *Niles Weekly Register* 12(May 31, 1817): 222; (St. Louis) *Missouri Gazette,* June 7, 1817; *Lexington* (Kentucky) *Reporter,* May 7, 1817.

31. Bierck, "John Hamilton Robinson," pp. 666-67; Foure Selter, *L'Odyssee,* pp. 77-80, 87-93.

32. Bierck, "John Hamilton Robinson," p. 667; Foure Selter, *L'Odyssee,* pp. 83, 96; (St. Louis) *Missouri Gazette,* March 27, 1818.

33. ibid.

34. Foure Selter, *L'Odyssee,* pp. 98-103; Hamilton E. V. Robinson to Antoine Saugrain, December 26, 1818, Saugrain Papers, Missouri Historical Society; Robinson to the Museum Society of Cincinnati, January 21, 1819, Gratz Collection, Historical Society of Pennsylvania.

35. John Hamilton Robinson, *A Map of Mexico, Louisiana, and the Missouri Territory, including the State of Mississippi, Alabama Territory, East & West Florida, Georgia, South Carolina & Part of the Island of Cuba* (Philadelphia, 1819).

36. David McNeely Stauffer, *American Engravers upon Copper and Steel* (New York: The Grolier Club, 1907), p. 9; Mantle Fielding, *American Engravers upon Copper and Steel* (Philadelphia, 1917), pp. 50-51; George C. Groce and David H. Wallace, *The New York Historical Society's Dictionary of Artists in America, 1564-1860* (New Haven: Yale University Press,

1957), p. 8; H. Glenn Brown and Maude O. Brown, "A Directory of the Book-Arts and Book Trade in Philadelphia to 1820, including Printers and Engravers," *Bulletin of the New York Public Library* 53(1949): pp. 221, 621.

37. Jackson, *Journals of Pike,* 1:413, 414,420-22; 2:64, 147, 153, 158, 178, 190-91, 205, 206, 237, 239.

38. Juan Pedro Walker, ["Map of central and western part of North America, from Lake Huron to the Pacific Coast"], Huntington Library, HM 2048; Wheat, *Transmississippi West,* 2:64-66, number 325.

39. Alexander von Humboldt, *Political Essay on the Kingdom of New Spain,* 4 vols., (New York: I. Riley, 1811), 1: xxxv, liii, liv.

40. Wheat, *Transmississippi West,* 2:71.

41. Zebulon Montgomery Pike, *A Chart of the Internal Part of Louisiana, Including all the hitherto unexplored countries . . .* published in Pike, *An Account of Expeditions to the Sources of the Mississippi, and through the Western Parts of Louisiana . . .* (Philadelphia: C. & A. Conrad, 1810).

42. See for example Wheat, *Transmississippi West,* 2:69-73, number 334; Thomas W. Streeter, *Bibliography of Texas, 1795-1845,* 5 vols., (Cambridge, Mass.: Harvard University Press, 1955-60), Pt. 3, vol. 1, pp. 53-55, number 1073.

43. Quoted in Wheat, *Tranmississippi West,* 2:72, note 13.

44. Foure Selter, *L'Odyssee,* pp. 98-103.

45. The best discussion of these negotiations remains Marshall, *Louisiana Purchase,* see especially pp. 53-66.

46. See, for example, note 48 below.

47. *Niles Weekly Register* 14 (July 18, 1818): 359, emphasis in the original.

48. Reprinted in *Niles Weekly Register* 15 (August 29, 1818): 6.

49. Streeter lists seven copies of the map: Library of Congress, Harvard, Princeton, New York Public Library, American Geographical Society, State Historical Society of Wisconsin, and his own personal copy, now at Yale. The National Union Catalog lists one copy at the New Hampshire State Library. There is also

a copy in the Cartographic History Library, the University of Texas at Arlington. The author has examined the copies at the Library of Congress, Wisconsin State Historical Society, New York Public Library and the University of Texas at Arlington. In addition, he has received reliable reports from American Geographical Society, and Yale. It is upon these examinations that the bibliographical analysis described is based.

50. The Library of Congress copy is of this state.
51. The University of Texas at Arlington copy is of this state.
52. The State Historical Society of Wisconsin and the New York Public Library copy is of this state.
53. Warren, *Sword,* 233-54.
54. (St. Louis) *Missouri Gazette,* November 24, 1819.

A TALE OF TWO CARTOGRAPHERS: EMORY, WARREN, AND THEIR MAPS OF THE TRANS-MISSISSIPPI WEST

Frank N. Schubert
Office of the Chief of Engineers
U.S. Army

Shortly before the Civil War,

the United States government published two important maps of the trans-Mississippi West. The cartographers were distinguished officers of the Corps of Topographical Engineers. Red-whiskered and quick-tempered Major William H. Emory (1811-1887), who compiled his map for the Office of the Mexican Boundary Survey, was a veteran explorer. He had conducted a major southwestern reconnaissance while attached to General Stephen W. Kearney's Army of the West during the Mexican War and spent several arduous years with field parties on the boundary survey. Lieutenant Gouverneur K. Warren (1830-1882), lean, intense, and at twenty-eight, nineteen years younger than Emory, published his map for the Office of Pacific Railroad Surveys. Warren had already worked on the Mississippi Delta survey and commanded two exploring expeditions into Nebraska and Dakota.

Scholars agree that the appearance of these maps culminated a half century of government exploration. Slowly at first and then in a burst of energy that accompanied the expansion of the 1840's and '50's, explorers crisscrossed the West. Lewis and Clark's trek across the continent was followed by Zebulon Pike's reconnaissances of the Southwest and the upper Mississippi. In 1819 Stephen Long crossed the plains to the Rocky Mountains via the Missouri and the Platte. During the next twenty years, government explorers, including Long, Lewis Cass, Henry Schoolcraft, and Joseph Nicollet, probed the huge triangle between the Mississippi, the Missouri, and British Canada. In the 1840's John C. Fremont, the flamboyant pathfinder, showed the way to Oregon and California. Other Topographical Engineers, among them Emory, James Abert, George Derby, Howard Stansbury, and William Simpson, led exploring parties to the Southwest. Then in the mid-1850's the Pacific railroad surveys added immense amounts of information to the bits and pieces gathered over the years. Finally, just a few years before war rent the nation, Emory and Warren fit the fragments into a reasonably correct mosaic and the nation had its first accurate overall picture of the region. The maps marked the end of the period of reconnaissance.

Fig. 8. Detail from Emory's "Map of the United States and their Territories Between the Mississippi and the Pacific Ocean and Part of Mexico."

While there is agreement on the importance of the government's 1858 maps there is some dispute over which of the two was more consequential. Most historians emphasize Warren's achievement. William H. Goetzmann, for example, noted Emory's map as representing an important step forward although it was soon superseded by Warren's. Goetzmann considered the publication of Warren's map as no less important than the appearance of the reports of Lewis and Clark.[1] Carl Wheat also rated Warren's as more important than Emory's.[2] The most recent narrative, John Noble Wilford's *The Mapmakers,* ignored Emory's achievement and concentrated on Warren.[3]

Emory has his defenders. Ralph Ehrenberg's 1971 paper before the Society for the History of Discoveries recalled Emory's significant role in developing the methodology used on both maps and in actually starting the Warren map while in charge of the Office of Pacific Railroad Surveys when it opened in 1854.[4] More recently, Martha Bray went all the way over to Emory, crediting him with the first reasonably correct map of the West, a map, she said, that climaxed fifty years of western exploration. She never even mentioned Warren.[5]

If we presume to ask what Warren and Emory might say about all of this, we might expect them to shout in unison, "I did it." Emory was the more temperamental of the two, but Warren was far from retiring. Neither was bashful about touting his own accomplishments, and both were somewhat touchy about any slights of their achievements.

Emory's temper and self-esteem were well-known long before his involvement with the map. As far back as 1849, when he unhappily joined the boundary survey in a position apparently subordinate to civilian astronomer Andrew Gray, Emory's commander, Colonel John J. Abert of the Corps of Topographical Engineers, had cautioned him to "avoid getting upon stilts."[6] Two years later, Emory bristled at a newspaper report that identified him as assistant to John R. Bartlett, the civilian commissioner. "I have seen," he wrote, "rather too much service to be an Assistant to any man that has not a pair of Epaulettes on his shoulders and who does not rank me in the Army."[7]

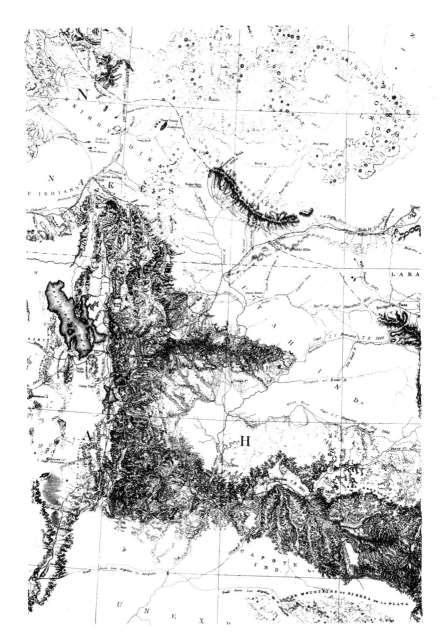

Fig. 9. Great Salt Lake Region detail from Warren's "Map of the
        Territory of the United States from the Mississippi to the
        Pacific Ocean."
        Courtesy of the University of Illinois, Map and Geography
        Library.

Emory's vanity was well known. Spencer Baird, the Smithsonian Institution's assistant secretary, played on Emory's pride like a skilled musician. Repeatedly Baird praised his contributions to the museum's natural history collections. When specimens came too slowly, Baird urged haste by reminding Emory that competitors would get the credit. You must hurry, Baird once admonished, if you will have "the honor of discovery."[8]

Twenty years later, Emory still had his pride and still read the newspapers. In response to an 1874 article that ignored his accomplishments as an explorer and cartographer, he again came to his own defense. Infuriated less by the work of a "fugitive newspaper writer" than by an Engineer officer's role in providing information he deemed inaccurate, Emory detailed his accomplishments. He claimed that he had first projected the general map and, based on the concept of Joseph Nicollet, had devised the plan by which the map was made. Moreover, he had obtained War Department approval for its production.[9]

Warren showed just as much interest in his own reputation. Even while he worked on his map and Emory finished his, Warren exhorted engraver Selmar Siebert to hurry. Siebert's delays meant that "our map has probably been set aside till the completion of the reduced copy from it engraved for the Mexican Boundary office."[10] Warren not only feared that Emory would beat him into print; he came within an inch of accusing his rival of plagiarism.

In later years, Warren still showed a great deal of sensitivity regarding his reputation. His summary removal from command of a Corps a few days before Lee surrendered at Appomattox, when he was only thirty-five but wore the two stars of a volunteer major general, certainly contributed to his touchiness, but General Philip Sheridan's harsh action against him stemmed more from Warren's willingness to publicly criticize the actions of his superiors and elders than from any alleged blunder.[11] Subsequently, Warren protested the failure of later mapmakers to credit him with discoveries he considered his.[12] He even got into a squabble over the dates of his brevet promotions, honorific appointments that bestowed neither permanent rank nor pay increases.[13]

Efforts at self-defense and self-promotion were not unusual among mapmakers of the day. Warren himself noted the secretive and competitive nature of government cartography. When he needed data in the hands of the Indian Bureau of the Interior Department, he found the custodians of the information reluctant to share it. As he wrote to Captain Andrew A. Humphreys, "My experience teaches me that as soon as it is known in Washington that one bureau is compiling a map all the others strive to withhold their information (always excepting the Land Office) so I doubt whether the Indian Bureau map can be obtained before it is published."[14]

With both Emory and Warren emphasizing their own roles and historians crediting one or the other with the major achievement, we have lost sight of the corporate character of the accomplishment. It is unlikely that Emory and Warren toiled in harmony, but they did work toward the same goal. Their efforts and their contributions overlapped and enhanced the net result. Several aspects of the production of their maps indicate that the venture was joint, if not cooperative.

First was the administrative organization of the enterprise. When Warren first reported to the Office of Pacific Railroad Surveys to compile the map, Emory was in charge of the office. Emory soon left to complete the survey of the Gadsden Purchase, but he returned to Washington to prepare his report and map in the Office of the Mexican Boundary Survey. As Warren's supervisor, he had the opportunity to influence development of the methodology; as Warren's competitor, he had access to his data.

More important is the evolution of the map itself. Emory did indeed play a major role in establishing standards for the map, the first of the entire West that contained no conjecture. He also contributed much of the material on which the southwestern portion was based. Warren too exerted a major influence on its development. He executed the final product, an endeavor that required him to be as much historian as mapmaker. In the course of this effort, he tracked down the reports, maps, and notes of many expeditions, and assessed the validity of their findings and the methods used to derive them. He also recommended a series of explorations to fill in the blank spaces along the Colorado and upper Yellowstone, in central

Nebraska and in the Black Hills. These surveys were completed by 1861, and Warren himself led the Nebraska and Black Hills expedition, which was his third. When he finished the map, he wrote an essay that explained the method behind the compilation, without, it should be noted, crediting Emory for any of it.[15]

The actual products reflected the intermingling of talent and labor. The first map, Warren's 1855 rendition of the routes followed by the Pacific railroad expeditions, was a preliminary effort.[16] As Warren noted on the margin, he considered the map a hurried compilation of authentic surveys that showed the relationship of the routes to each other. Emory's map of 1858, drawn to the same 1:6,000,000 scale as Warren's earlier one and engraved by the same Selmar Siebert who did Warren's later map, contained much more detail. It particularly improved on Warren's depiction of the mountains.[17] Warren in 1855 had shown only the railroad passes; Emory included the entire ranges. Warren in 1858 added significant material, particularly to the interior of Oregon. A second edition that appeared soon after the first also reflected Warren's third expedition and Joseph C. Ives's 1858 examination of the Grand Canyon.[18] Neither edition made any noticeable corrections to material included in Emory's.

The significance of the map we have come to call Warren's went beyond merely symbolizing the end of the period of reconnaissance. Warren's map went through numerous editions. After the Civil War an inset map of Alaska was included. For a generation, it remained the basic map of the western states and territories.

Ultimately the question of the relative importance of Emory's and Warren's contributions obscures the larger fact of their joint participation in a major enterprise. Their efforts, built on those of numerous explorers, sustained by government agencies, and financed by the nation's taxpayers, came together in Washington just before the Civil War. Almost on the eve of the great coming-apart, they drew so much together. For the first time, Americans could turn to a map — call it Emory's or call it Warren's — and see the main features of the huge western domain.

NOTES

1. William H. Goetzmann, *Army Exploration in the American West 1803-1863* (New Haven: Yale University Press, 1959), pp. 200, 313.

2. Carl I. Wheat, "Mapping the American West," *Proceedings of the American Antiquarian Society,* 64 (April 1954), pp. 160-61, 163.

3. John Noble Wilford, *The Mapmakers* (New York: Alfred A. Knopf, 1981), pp. 202-205.

4. Ralph E. Ehrenberg, "Exploring the Trans-Mississippi West: The War Department's Office of Explorations and Surveys, 1857-1861," unpublished paper read at the annual meeting of the Society for the History of Discoveries, November 12, 1971, pp. 3-4.

5. Martha C. Bray, *Joseph Nicollet and his Map* (Philadephia: The American Philosophical Society, 1980), p. 270.

6. Colonel John J. Abert to Emory, 29 May 1849, Emory Papers, Yale University.

7. Emory to Captain [?], 28 September 1851, Emory Papers.

8. Spencer F. Baird to Emory, 10 January 1853, Spencer F. Baird Papers, Private Correspondence, Volume II, Smithsonian Institution Archives.

9. Emory to George M. Wheeler, 21 May 1874, Letters Received, General Records Division, Office of the Chief of Engineers, 1871-1886, Box 15, Record Group 77, National Archives.

10. Warren to Selmar Siebert (pencil draft), 3 February 1858, Correspondence of the Office of Explorations and Surveys, Box 1, Binder 1, Record Group 48, National Archives.

11. See Louis H. Manarin, "Major-General Gouverneur Kemble Warren: A Reappraisal," unpublished M.A. thesis, Duke University, 1957.

12. Warren to Brigadier General A. A. Humphreys, 11 November 1874, Letters Received, General Records Division, Office of the Chief of Engineers, 1871-1886, Box 15.

13. Warren to Secretary of War William W. Belknap, 8 February 1876, Adjutant General E. D. Townsend to Secretary of War Belknap, 18 February 1876, and Townsend to Warren, 21 February 1876, File 649 ACP 1876, Record Group 94, National Archives.

14. Warren to Humphreys, 22 October 1860, Correspondence of the Office of Explorations and Surveys, Box 4, Binder 1.

15. Gouverneur K. Warren, *Memoir to Accompany the Map of the Territory of the United States from the Mississippi River to the Pacific Ocean,* 33d Congress, 2d sess., Senate Executive Document 78, Volume XI (Serial Set No. 801).

16. Gouverneur K. Warren, "Map of Routes for a Pacific Railroad Compiled to accompany the Report of the Hon. Jefferson Davis, Sec. of War, in Office of P.R.R. Surveys, 1855," filed under "United States-West-Exploration" in the Geography and Map Division, Library of Congress.

17. Emory's "Map of the United States and their Territories Between the Mississippi and the Pacific Ocean and Part of Mexico," is G4050 1858.J4, Geography and Map Division.

18. The first edition of Warren's "Map of the Territory of the United States from the Mississippi to the Pacific Ocean" (classified as G4050 1857 .W32 by the Geography and Map Division) is dated 1857. However, an incomplete proof of the map (G4050 1857 .W35) bears Warren's signature and the date of 8 January 1858. The second edition, dated 1858, includes material from Joseph C. Ives' 1858 survey of the Colorado River and was sent to the printer in October, 1859. Warren to the Superintendent of Public Printing, 26 October 1859, Correspondence of the Office of Explorations and Surveys, Box 4, Binder 1.

EXPLORATION AND MAPPING OF THE SOUTHWEST ROUTE, FROM MISSOURI TO SOUTHERN CALIFORNIA

James A. Coombs

Southwest Missouri State University
Springfield

The first people to explore

routes in the area of this study were native Americans of the Rio Grande valley and California coast following the migration trails of animals. If there are written accounts of their explorations or maps of these trails still in existence today, scholars have yet to find them. The first European explorers in the American southwest found that the natives apparently conducted a regular trade in seashells, deer skins, and blankets, and they benefited immeasurably from the knowledge of the native American traders who guided them over these trade routes.

SPANISH EXPLORATIONS IN NEW SPAIN, 1600 — 1800

In 16th Century New Spain, the tales of Coronado's and Antonio de Espejo's expeditions to the north led to numerous proposals to relocate and find gold in legendary Quivira. Apparently, the earliest map drawn from actual reconnaissance of any part of the area concerned in this study resulted from one of these explorations. In 1601, Don Juan de Oñate led a party from San Juan, New Mexico, northwest of Santa Fe, to the northeast in search of Quivira. Their route went from San Juan east across the Pecos River to the Canadian, then downstream 111 leagues. From there, the expedition went northeast across the Arkansas River to Quivira, a Wichita Indian village about 25 miles southeast of present-day Wichita, Kansas.[1]

The map of Oñate's exploration was drawn by Enrique Martínez, the Cosmographer to the King of Spain. Martínez was not on the trip, but sketched the map from a description by Juan Rodríguez, a sailor who accompanied Oñate. At first glance, this map looks fairly crude. When compared to other maps of the same area made at this time, however, one can appreciate its achievement. Although east-west distances are greatly foreshortened, the rivers and tributaries of the southern Great Plains are correctly shown. Most importantly, the Rio Grande is shown emptying into the Gulf of Mexico, rather than the Gulf of California, an inaccuracy other 17th century European cartographers included on their maps.[2]

In 1604, Oñate led an expedition west from San Juan, past the Zuni and Hopi settlements, southwest across the Little Colorado and Verde rivers, then west to the Bill Williams Fork, which they descended to the Colorado. After reaching the mouth of the Colorado, the party returned to New Mexico by the same route.[3] They stopped at the base of a sandstone monolith now called El Morro, and Oñate carved the earliest still-decipherable inscription: "There passed this way Adelantado Don Juan de Oñate from the discovering of the South Sea, on the 16th of April, 1605."[4] El Morro would be a stopping place for many explorers through the centuries. It was unique among the many flat-topped hills in western New Mexico because of a pool of fresh water at its base. It became the objective of many travelers, after spending a relatively comfortable night in this arid wilderness, to record their presence before setting out into the unknown again.

Oñate's explorations are significant to this study because direct routes were found from the Rio Grande valley to the Colorado River and to the eastern Great Plains. To Oñate's superiors and to the members of his expeditions, however, they were failures since no gold or jewels were found. Due to this, and to the insurrections of the mistreated natives in New Mexico, Spain neglected its northern settlements for the rest of the 17th and the first three-quarters of the 18th centuries. Encroachments from the north and east by other European countries, however, eventually motivated the Viceroy of New Spain to support and supply the northern settlements and to establish an overland road between outposts in New Mexico, Alta Pimeria, and Alta California.

Two notable expeditions took place in 1776 which did more to establish the feasibility of land routes between the Rio Grande valley and Monterey Bay than any others. The first was conducted by Francisco Tomas Hermenegildo Garcés, who, in February 1776, while visiting the Mojave villages on the Colorado River, near present-day Needles, learned that the Mojaves frequently traveled west to the Spanish missions on the California coast. Garcés recruited some Mojaves as guides who led him to the San Gabriel mission, in present-day Los Angeles, via the Mojave River and Cajon Pass through the San Bernadino Mountains.

After returning to the Mojave villages in May, Garcés recruited some visiting Hualapai Indians to guide him east to the Hopi lands. They headed northeast to the Grand Canyon village of the Havasupais at the bottom of Cataract Canyon, then went on to Oraibe, the Hopi village in northeastern Arizona. Garcés sent a letter announcing his arrival in Oraibe to the resident padre at Zuni, Francisco Silvestre Velez de Escalante, and asked that it be forwarded to the Governor of New Mexico, in order to prove that a road between Alta California and New Mexico was feasible. A map of Garcés' travels does exist, but the extent of its use by other explorers and cartographers is not known.[5]

Father Escalante received Garcés' letter in Santa Fe, where he and Father Atanasio Domínguez were preparing an expedition of their own to find a route from Santa Fe to Monterey Bay. Escalante's belief that the topography west of Oraibe was inhospitable was not changed by Garcés' news, unfortunately, so this, the other notable expedition of 1776, headed *northwest* out of Santa Fe on July 29. They crossed the Uncompahgre Plateau, reaching the Colorado River near present-day Grand Junction, Colorado. The party continued northwest and west over the Wasatch Mountains to Utah Lake. Due to lack of supplies and fear of winter, Escalante and Domínguez decided there to abandon their attempt to reach Monterey Bay. They headed back to Santa Fe by traveling south to a point below the Hurricane Cliffs, just north of the Grand Canyon. Because of the rough terrain to the south, the group turned east, following the Colorado upstream until they found an easy crossing in southern Utah ever since called "The Crossing of the Fathers." They continued east from that point, visiting Oraibe, and eventually arriving back in Santa Fe on January 2, 1777.[6]

A cartographer on the expedition, Bernardo de Miera y Pacheco, drew a remarkable map to accompany the official expedition report. It is known today to exist in six manuscript copies, but it was apparently not published for more than a century after its creation. Even so, a few cartographers, such as Baron von Humboldt, had access to it in manuscript form and used its information on their own maps.[7]

The reports of Garcés and Escalante, as well as those of Fathers Pedro Font and Tomas Eixarch and Captain Juan Bautista de Anza, who also explored routes between Alta Pimería and Alta California, convinced the officialdom of New Spain to strengthen its influence in its northern frontier settlements. The commandant-general of the frontier provinces devised a plan to build small mission-pueblos at various intervals along supply routes, but it failed. The Yuma Indians, angered by settlers' intrusions on their lands, revolted and wiped out the only two colonies established.[8]

The territory north and east of New Mexico, meanwhile, had been little explored by the Spanish since Oñate, mostly due to the hostile Comanches, Apaches, and Pawnees living there. The Spanish were aware of British and French explorations and trading activity north of the Missouri River, but did not consider their New Mexico settlements threatened until French explorers such as the Mallet brothers reached Santa Fe.[9] Even then, the Spanish were not hostile toward the traders, merely uneasy. They considered the French-owned no-mans land between the Arkansas and Missouri rivers an adequate buffer zone. In 1763 the situation changed, however, when Spain acquired this land from France and its sovereignty moved to the west bank of the Mississippi River. The French were no longer a threat, but the British were more of a threat than ever.

To prevent British traders from entering Louisiana, as this vast territory was called, Spanish forces were sent to construct a fort at the confluence of the Missouri and Mississippi rivers and a lieutenant-governor's headquarters was established at the recently founded village of St. Louis. Even with this attempt to control the northern frontier, English traders carried on commerce with the Indians on both sides of the Missouri and Mississippi rivers and their tributaries.[10]

This situation prompted the Spanish to open a trail from Santa Fe to St. Louis to be used as a supply line to the northern border. The Viceroy of New Spain commissioned Pedro Vial, a well-known trailblazer of the time,[11] to establish this trail. Vial and two companions set out from Santa Fe on May 21, 1792, with instructions to "keep a diary . . . , marking in it the courses and daily distances, the rivers encountered, the mountains and tablelands . . . explaining their configuration."[12] On the way to St. Louis, the party was captured by Kansas Indians. They would have been killed had not one of the Indians recognized Vial. Even so, they were taken to the Kansas village on the Kansas River and held captive. Six weeks later, a French trader came into the village, rescued Vial and his men, and accompanied them to St. Louis. They stayed in St. Louis from October 3, 1792 until June 14, 1793. On the trek back to Santa Fe, Vial's group first visited the Pawnee villages in northeast Kansas, then headed southwest. They were later attacked by Pawnees who mistook them for Comanches. No loss of life resulted, however, and the expedition arrived back in Santa Fe November 16, 1793.[13]

There is no evidence that Vial drew a map of the country traversed on this trip to accompany his diary. He had drawn a remarkable map in 1787, however, covering all the territory in the trans-Mississippi west he had already traveled. It contains quite a bit of accurate information of the entire Mississippi-Missouri watershed, as well as some of the Rio Grande basin. There is no record of Vial ever traveling north of the Missouri, and little chance that he might have obtained cartographic information from the French or British explorers who did. Even in areas for which record of his travels exists, the accuracy of the courses and length of rivers and locations of places is very good considering Vial carried only a compass on his travels and could only estimate the distances traversed each day.[14]

After Vial stated that he could have made the trip from Santa Fe to St. Louis in twenty-five days if he had not been captured by Indians, the Spanish officials in the Interior Provinces abruptly realized how exposed and vulnerable Santa Fe was to the Anglo-Americans who were probing the rivers and plains in efforts to find that fabled key to the riches of the unknown west. Vial and others made several trips to the Pawnee villages in the next few years, but in spite of their efforts to keep the Indians friendly with New Spain, Spanish influence in the Louisiana Territory faltered toward the turn of the century. The King of Spain eventually ceded the Territory to Napoleon, who in turn sold it to the United States in 1804.[15]

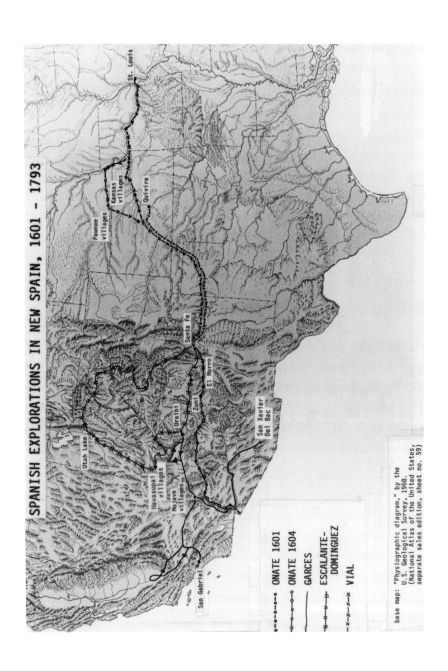

SPANISH EXPLORATIONS IN NEW SPAIN, 1601 – 1793

St. Louis

Kansas villages

Pawnee villages

Quivira

Santa Fe

El Morro

Oraibi

Zuni

San Xavier Del Bac

Utah Lake

Havasupai villages

Mojave villages

San Gabriel

ONATE 1601

ONATE 1604

GARCES

ESCALANTE-
DOMINGUEZ

VIAL

base map: "Physiographic diagram," by the
U.S. Geological Survey, 1968.
(National Atlas of the United States,
separate sales edition, sheet no. 59)

AMERICAN ROUTES TO SANTA FE 1806 - 1846

On March 10, 1804, the United States became the official owner of the Louisiana Territory. James Wilkinson, its first Governor-General, began to plan military expeditions to establish control. He directed Lt. Zebulon M. Pike to head a reconnaissance of the Plains Indians country between St. Louis and Santa Fe, to investigate the headwaters of the Arkansas and Red rivers, and possibly to establish a trail to Santa Fe for commerce. Whether Wilkinson sent Pike on this expedition with ulterior motives against the Spanish has been subject to controversy ever since.[16]

Pike's party, consisting of Dr. John Robinson, Lt. James B. Wilkinson (the Governor-General's son), nineteen other military personnel, and an interpreter, left St. Louis on July 15, 1806, while Lewis and Clark were still homeward bound in the far upper Missouri River area. For surveying and mapping, Pike took along a theodolite, telescope, sextant, chronometer, and compasses. He also brought a sketch map he drew from information provided by earlier travelers and possibly from Humboldt's map of New Spain. The geographic features shown on this map are not ones Pike's expedition observed, nor is the route shown his. Most authorities agree that the route shown is the Santa Fe Trail.[17]

The route of Pike's expedition can be followed on the published maps which accompany his journal.[18] The party first escorted fifty-one Osage Indians up the Missouri and Osage rivers to their villages, then headed northwest to the Republican Fork of the Kansas River to hold council with the Pawnees at their village. There Pike learned that Lt. Facundo Melgares had been sent from Santa Fe in April 1806 to intercept him, and had recently left the Pawnees to return to Santa Fe.

Leaving the Pawnees, the Americans went south to the Arkansas River, reaching it near Pawnee Rock. Here the party split up, with Lt. Wilkinson taking some men down the Arkansas to determine its navigability. Pike and the others, having found Melgare's trail, followed it upstream to the mouth of Fountain Creek. Apparently expecting to meet the Spanish at any time, Pike instructed his men to build a breastwork for protection while he and Dr. Robinson headed for the "Blue Mountain" (now Pike's Peak). Failing to reach the summit, they returned to the breastwork and the entire party headed upstream to the base of the mountains, at present-day Cañon City. Determined to follow what he believed to be Melgare's trail, Pike led his party up Four-Mile Creek to the South Platte River near present-day Elevenmile Canyon Reservoir. Here he realized he had been following a very stale Indian trail, so they turned west and came to the Arkansas River near present-day Buena Vista, Colorado. Pike thought they had come upon the Red River, and the group descended it, struggled through Royal Gorge, only to realize it was the Arkansas when they discovered their old campsite near Canon City.

Still determined to find the Red River, the expedition headed southwest up Grape Creek, across the Sangre de Cristo Mountains near the Great Sand Dunes National Monument, and on to the Rio Grande near present-day Alamosa, Colorado. By this time it was the middle of winter and the party was in dire straits. The men suffered from frozen feet and lack of food, so Pike decided to build a stockade on Conejos Creek, a western tributary of the Rio Grande in southern Colorado. Dr. Robinson then decided the time was right for him to go to Santa Fe, one hundred miles to the south, to attend to some pecuniary matters entrusted to him in 1804 by William Morrison, a Kaskaskia, Illinois merchant. The motivation for sending Robinson off alone to almost certain capture has been subject to much discussion, but let it suffice here to say that he left the stockade on February 7, 1807, and was taken prisoner by the Spanish in Santa Fe.[19]

On February 26, Spanish soldiers from Santa Fe came to Pike's stockade and informed him that Dr. Robinson had arrived in Santa Fe. When asked why he was in Spanish territory, Pike professed ignorance and told them he thought he was on the Red River, and in United States Territory. The Spaniards offered to supply Pike's expedition with mules, conduct them to the head of the Red, and send them on their way, *after* they went to Santa Fe and explained to the Governor of New Mexico their business on the Spanish frontier. In Santa Fe, however, the Governor found Pike's sketch map while examining his papers, and assumed it showed a planned U.S.

military route to Santa Fe. He then had the Americans taken to Chihuahua to appear before Nemesio Salcedo, the Commandant General of the Internal Provinces. From this point on, although they were treated with civility, Pike and his men were considered prisoners.

On the way to Chihuahua, in a small town just south of Albuquerque, the Americans were reunited with Dr. Robinson. He was also on his way to Chihuahua while touring the Internal Provinces, practicing medicine and observing the customs of the people. Lt. Melgares, who had been escorting Dr. Robinson to Chihuahua, finally met Pike here, and became his escort also. Also while in this area, at Father Ambrosio Guerra's house, Pike had his only opportunity to see a map of the Province of New Mexico, which he noted as depicting the sources of the Rio Grande and the Colorado River.

In Chihuahua, Pike shared quarters with Juan Pedro Walker, who assisted Andrew Ellicott in the 1796-1800 survey of the southern boundary of the U.S./New Spain border, and who now surveyed for the Spanish government.[20] Walker's maps of the provinces of New Spain were taken off the walls for Pike's stay, an indication of General Salcedo's suspicions of the Americans' motives in his country. The General and Walker examined Pike's papers as he explained each one, and he drew a sketch map for them of his concept of his travels since leaving St. Louis. The Americans stayed in Chihuahua for almost a month before they were allowed to leave on April 28. Much to Pike's dismay, the General decided to confiscate his papers. Fortunately, Pike had kept his field notes hidden all this time, and succeeded in getting them back to the United States intact. The Americans finally reached U.S. territory on June 30, 1807, when they rode into the fort at Natchitoches on the Red River in present-day Louisiana.[21]

Soon after his release from New Spain, Pike enlisted the help of Anthony Nau to draw the maps which accompany his *Account*. His route from St. Louis to Santa Fe is shown on the two-part map "The Internal Part of Louisiana," while the route from Santa Fe to Natchitoches is shown on "The Internal Provinces of New Spain."[22] Like most maps of that day, Pike's represent a mixture of cartographic plagiarism and original observation. The cartography on the eastern half of "The Internal Part of Louisiana" was taken from Lt. Wilkinson's survey of the lower Arkansas River, Abraham Bradley's 1804 "Map of the United States," and Nicholas King's maps of the Lewis and Clark, and Dunbar and Hunter expeditions. All that Pike contributed was his own route. The western half of this map has more historical significance because it was constructed from Pike's field notes. Cartographic information was lifted from Baron Humboldt's map of New Spain or is imaginary, so it turns out that his route is the most accurate feature on this sheet. "The Internal Provinces of New Spain" is, with a few changes and additions, literally a copy of Humboldt's manuscript map. As with the eastern half of the other map, all Pike added was his route. The responsibility for plagiarism more than likely lies with Nau. Although Pike helped him with the rough drafts or manuscripts, Nau drew the published versions on his own, and apparently borrowed information from the other maps to fill in the blank spots.[23]

Despite the plagiarism and imaginary cartography, as well as his many errors and misconceptions, Pike's journal and maps, published in 1810, provided the best account of exploration in the southwest that the American public had available. The information on his maps was used by other cartographers, most notably by Juan Pedro Walker and Dr. John Robinson. Walker's map, an untitled, undated manuscript of what is now the United States west of the Mississippi River, contains some features in the New Mexico area which are identical to those on Pike's maps. Walker undoubtedly used the sketch maps Pike drew for General Salcedo in Chihuahua, and possibly his published maps as well, when drawing this manuscript map.[24] Dr. Robinson's map is a quite large (64 x 67 inches) depiction of "Mexico, Louisiana, and the Missouri Territory, including also the state of Mississippi, Alabama Territory, East & West Florida, Georgia, South Carolina & part of the island of Cuba," which he drew in 1819. Not only is this map notable because it reflects Pike's cartography, but also because it contains information from Robinson's own experiences, the routes of Escalante, Font and Garcés, information from Walker's maps, and because it was the first to name Pike's Peak.[25]

Pike's journal and maps, as well as early traders' reports of fantastic profits reaped in Santa Fe, were instrumental in beginning the Santa Fe trade. Perhaps the most colorful account of those early expeditions is that of William Becknell, who in 1822 was the first to take wagons and head out across the unknown desert south of the Arkansas River. With no other guide than the stars and possibly a compass, Becknell's party lost its way and almost perished from lack of food and water. Frantic with despair, they scattered in all directions, not realizing they were almost upon the Cimarron River. Some of the men found a buffalo with its stomach distended with water from the river. They killed it and quenched their thirst with the "invigorating draught procured from its stomach."[26] Soon all the men had recovered and the journey resumed. Becknell's route became known as the Cimarron Cutoff.[27]

A combination of the increase in the volume of traffic and the Indian molestations on the Trail led to a call in Congress for protection of traders and an official route survey. In 1825, the Kansas-Osage Treaty was signed, and three U.S. Commissioners, Benjamin H. Reeves, George C. Sibley, and Thomas Mather, along with surveyor Joseph C. Brown, made the first actual survey of the Trail, or any part of the West, for that matter. They surveyed a route from Ft. Osage, near present-day Independence, Missouri, to Taos with a surveyor's chain, compass, and sextant, and marked it with mounds of earth or stone. Brown compiled a "chart and way bill" made up of section maps from camp to camp with remarks about notable places, and used it later to construct a route book and a map. He apparently intended to publish these in 1827, but it never came to be. The manuscripts were apparently used by the U.S. Bureau of Topographic Engineers and by Joshua Gregg when compiling his map of the prairies.[28]

Despite the U.S. Government's efforts to stabilize the Santa Fe Trail, many traders continued to blaze their own routes. Joshua Gregg, for example, left Van Buren, Arkansas on April 21, 1939, and traveled up the north side of the Canadian River, reaching Santa Fe on June 25. He navigated by compass, sextant, and a map of the prairies drawn by a Comanche chief, which Gregg claimed was more accurate than any published map he had seen. The return trip was made the following Spring, leaving Santa Fe on February 25, 1840, traveling along the south side of the Canadian, and arriving in Van Buren on April 22. Gregg preferred this route to the established Santa Fe Trail because it was "some days" travel shorter, less intersected with large streams, fewer sandy stretches, . . . and grass springs up a month earlier than in Missouri."[29]

Soon after this trip, Gregg compiled the two-volume *Commerce of the Prairies: or the Journal of a Santa Fe Trader . . ."* [30] This classic description of commerce and travel in the American southwest before the advent of the railroad was first published in 1844, and was based on Gregg's own experiences during four of his eight crossings of the prairies. The first edition of this book contained his "Map of the Indian Territory, Northern Texas and New Mexico showing the Great Western Prairies," the most complete and reliable map of the area then in existence. It was based on his own field notes and records, as well as Humboldt's map of New Spain, Major Stephen Long's 1819 expedition map, and J.C. Brown's Sante Fe Trail survey.[31]

In 1845, the United States annexed Texas, and previously-strained U.S.-Mexican relations rapidly deteriorated. President Polk believed that military force would be necessary to maintain the new border with Mexico, but military strategists found the only maps of the region were Gregg's map of the prairies, H.S. Tanner's 1839 edition *Map of Mexico,* and Lt. William H. Emory's 1844 *Map of Texas and the country adjacent.*[32] These were found to lack the detail needed for military planning, so the U.S. Bureau of Topographic Engineers was instructed to conduct trans-Mississippi expeditions to gather geographic information. Captain John Fremont received orders in 1845 to conduct one of these expeditions, from Bent's Fort in southeastern Colorado to Ft. Gibson in eastern Oklahoma, via the Canadian River. Fremont delegated this survey to Lts. James W. Abert and William G. Peck, however, while he led an unauthorized expedition to California. Equipped with only a sextant and chronometer (Fremont took the other instruments), Abert and Peck left Bent's Fort on August 12, 1845, and gathered flora, fauna, and other geographical information along the south side of the Canadian. They proceeded downstream as rapidly as possible, for Indians continually

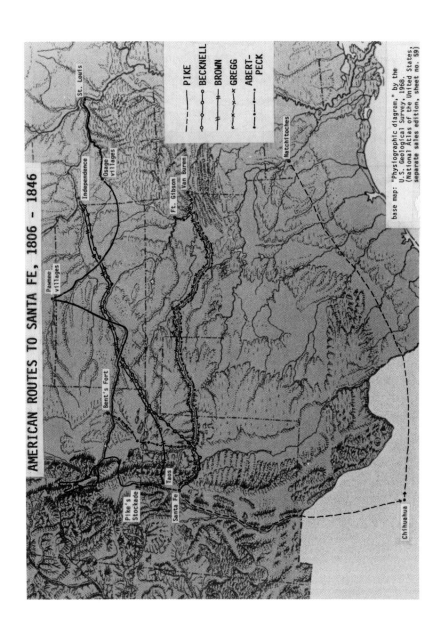

AMERICAN ROUTES TO SANTA FE, 1806 - 1846

PIKE
BECKNELL
BROWN
GREGG
ABERT-
PECK

St. Louis
Independence
Osage villages
Pawnee villages
Bent's Fort
Pike's Stockade
Taos
Santa Fe
Ft. Gibson
Van Buren
Natchitoches
Chihuahua

base map: "Physiographic diagram," by the
U.S. Geological Survey, 1968.
(National Atlas of the United States,
separate sales edition, sheet no. 59)

observed them from crags and promontories. The party was never attacked, fortunately, and they reached Ft. Gibson on October 21. Despite their hasty trip, Abert produced a carefully drawn map, which along with his report, provided authoritative information about the north Texas country and Canadian River region, and was used in constructing other Topographic Engineer maps.[33]

EXPLORATIONS OF ROUTES TO SOUTHERN CALIFORNIA, 1846 — 1854

In 1846, the United States went to war with Mexico for a number of reasons, not the least of which was a widely-held desire to acquire California. In June of that year, President Polk ordered Col. Stephen W. Kearney to invade New Mexico with his Army of the West, and to march on to California in order to establish a communication and supply line.[34]

The routes from Missouri to Santa Fe were established and well-mapped by this time, but the Army lacked accurate reports or maps of the area west of the Rio Grande. Therefore, an attachment of Topographic Engineers, led by 1st Lt. William H. Emory, was assigned to accompany Kearney's Army of the West. Lts. Abert and Peck were assigned to assist Lt. Emory with the topographic sketches, and two civilians, Norman Bestor and Prof. J.C. Hubbard, were recruited to help with the astronomical observations.

Emory was in Washington, D.C. when he received his orders, and he was given just twenty-four hours to gather scientific equipment. Though unable to obtain a pocket chronometer or telescope as he wanted, he did manage to get two box chronometers, two sextants, and a siphon barometer. He also took Tanner's 1846 map of Mexico, Mitchell's 1846 map of Texas, Oregon, and California, his own map of Texas, Lt. Abert's map of the Canadian River country, Capt. Fremont's report of his 1843-44 expedition to California, and Gregg's *Commerce of the Prairies*.

The Army of the West left Ft. Leavenworth on June 28, 1846, and had a relatively uneventful march along the Santa Fe Trail to Bent's Fort, arriving there July 29. Intelligence reports and rumors gave conflicting predictions of what lie ahead, but it appeared the

Mexicans in Sante Fe were prepared to fight. Kearney ordered Col. Philip St. George Cooke and James Magoffin to go there under a flag of truce to try to convince Gov. Manuel Armijo that resistance was futile. Magoffin was a long-time Santa Fe trader and Gov. Armijo's cousin by marriage, and was recruited specially by President Polk to personally negotiate with Armijo. How successful the Cooke-Magoffin mission was in persuading the Mexians not to fight is speculative, as no official or first-hand record of the conversations exist.[35] Armijo was evidently convinced of the futility of fighting, though, because his forces withdrew as the Army of the West advanced, and Sante Fe was occupied on August 18 without a shot being fired. Nevertheless, the march from Bent's Fort to Sante Fe was hard, as water and grass were scarce, and the soldiers were hungry most of the time due to supply problems.

The Army of the West occupied Santa Fe for 38 days. Emory's Topographic Engineers spent most of their time selecting and surveying the site of Ft. Marcy, constructing a map of the route followed from Ft. Leavenworth to Santa Fe, and making observations on contemporary life in Santa Fe for the official report.[36] On September 25, Kearney and a force of 300 dragoons began the trek to Califorinia by heading south out of Santa Fe along the Rio Grande. They were accompanied by Emory's mapping unit, which now included Lt. William H. Warner and artist John Mix Stanley. Lts. Abert and Peck had become ill and were instructed to stay behind and compile a map of the entire province of New Mexico.[37]

A few miles south of present-day Socorro, New Mexico, on October 6, the army met a party of about fifteen men led by Lt. Kit Carson, who was carrying sealed dispatches from California to Washington. Carson told them California had fallen without a blow to the combined forces of Capt. Robert F. Stockton and Lt.-Col. Fremont, and was now a possession of the United States. In light of this new development, Kearney reorganized and reduced his force to about 100 men. Carson, despite his protests, was pressed into service as a guide. He had been directed to deliver the dispatches personally, and realizing the honor that would be given to the man who notified the President that California was now an American possession, he wanted very much to carry out his orders. But to Kearney, Carson's assistance as a guide would be invaluable, since he had just traveled over the trail to California.

On October 15, near present-day Truth or Consequences, the army turned away from the Rio Grande and headed southwest over the Mimbres Mountains on the road to the Santa Rita copper mines. After a brief inspection of the mines, they marched west toward the Gila River, reaching it at the north end of the Big Burro Mountains. The trek down the Gila was very difficult, as the canyon walls were nearly at the water's edge and the expedition had to continually cross the river. In several places the gullies were so deep that they were forced away from the river into the mountains. Contacts with Apache traders were frequent, and the soldiers were glad to get fresh horses and mules, as the brutal labor of the trail was especially hard on the animals.

The army finally emerged onto flat country November 9, and wound its way down the Gila valley to the Pima villages. They stayed there a few days to rest and replenish their provisions, then set out again down the Gila, through a sandy country interspersed with promontories and practically devoid of vegetation. By the time the expedition reached the Colorado River, most of the men were afoot. The animals had subsisted mostly on cane and mesquite and were on the verge of collapse. Despite the hardships and suffering around him, Emory was absorbed with the strangeness of the country. He made extensive observations on the vegetation, animals, insects, geology, minerals, and archeology, as well as the astronomical observations necesssary for constructing maps.

Near the confluence of the Gila and Colorado rivers, the army came upon a small band of Mexicans driving a herd of horses to Sonora for the Mexican army's use. Questioned about the situation in California, the Mexicans replied that Santa Barbara and Los Angeles were now under Mexican control, while the Americans still controlled San Diego. This news was confirmed when a lone Mexican rider was brought into camp and searched. It turned out he was a courier with letters from California to General José Castro, which contained detailed information concerning the Mexican counter revolution. Kearney was now anxious to get to the California coast as quickly as possible, so the army picked fresh mounts from the Mexican herd, and some of the Mexicans were retained to act as guides.

On November 25, the Army of the West crossed the Colorado River near present-day Yuma, Arizona and entered the Colorado Desert (now the Imperial Valley). At the end of the three-day jornada, the men and animals were suffering severely, but they found a spring on Carrizo Creek at the base of the Tierra Blanca Mountains, about 60 miles east of San Diego, where there was plenty of water and cane for the animals. Though the desert was now behind them, their suffering was not over. From the spring, they headed northwest to Jonathan Warner's ranch, through barren, rugged country. Their provisions were almost gone and the animals gave out by the dozens. The weather, varying from hot, dry days to cold, foggy nights, added to their fatigue. This inhospitable land was not the California the Americans expected. Only with extreme difficulty was Gen. Kearney able to bring his army to Warner's ranch intact on December 2.

Meanwhile, the Californians under General José M. Flores learned that Kearney's force was at Warner's ranch, blocking their supply line from Sonora. Flores dispatched Don Andreas Picó and several squadrons of mounted lancers to regain control of the supply line. Gen. Kearney, learning of this, sent a message to San Diego for reinforcements, which came in the form of 37 men led by Captain Archibald Gillespie. The American and Mexican forces met at San Pascual, about 30 miles northwest of San Diego. After the initial clash of arms, in which eighteen Americans were killed and thirteen wounded while two Mexicans were killed and several wounded, the Americans were forced to a hilltop. Here, after a march of over five months and 1,887 miles, they nursed the wounded and subsisted on animal flesh while Kit Carson, Lt. Edward Beale, and an Indian scout slipped through to San Diego and returned with 200 reinforcements. The Army of the West was saved, and arrived in San Diego on December 12, 1846.[38]

The American forces soon left for Los Angeles, and Emory accompanied them. Messrs. Bestor and Stanley took the surveying instruments and logbooks on to a ship in the San Diego harbor to complete the compilations and construct a map. Published with Emory's report in 1848, it shows Col. Cooke's route as well as

Kearney's, and was the first accurately drawn of the area along the 32nd parallel. It corrected the errors in this area on the maps of Fremont, Humboldt, Tanner, and Mitchell. Although the Topographic Engineers' reconnaissance was rapid, predominated by military objectives, they did manage to make 2,000 astronomical observations, 357 barometric altitude observations, and 52 established points of latitude and longitude. Compiled from these, Emory's map remains one of the major contributions to western American cartography.[39]

Although Emory considered the Gila route along the 32nd parallel practical for the transcontinental railroad, it never became popular with later travelers. The Butterfield Overland Stage and the Southern Pacific Railroad did eventually follow the wagon road Col. Cooke's Mormon Battalion established along a more southern route around the southern end of the Mimbres Mountains of New Mexico, across the dry lake beds to Guadalupe Pass, and northwest to Tucson and the Gila River, however.[40]

There was soon a new motivation for exploration of routes to California: the gold rush of 1848-49. The best overland route was the topic of many discussions, and frontier citizens in Texas, Arkansas, and Missouri tried hard to promote trails that originated in their town. In one of these efforts, U.S. Senator Solon Borland of Arkansas petitioned the Secretary of War to have a survey conducted of the Ft. Smith-Santa Fe route. Accordingly, Capt. Randolph B. Marcy, with three companies of troops and Lt. James H. Simpson of the Topographic Engineers, escorted a large group of California goldseekers while surveying and constructing a road. They left Ft. Smith on April 4, 1849, and went up the south side of the Canadian River to Tucumcari, then west to Anton Chico on the Pecos River, and on up the established road to Santa Fe. Lt. Abert's 1845 map and report were of great help to Simpson, as Abert had surveyed approximately the same route on his hasty descent of the Canadian. This time, however, mileage was measured by both viameter and surveyor's chain, with compass bearings every mile. Both Marcy and Simpson compiled official reports and maps of their survey, and both enthusiastically recommended this route for the location of the transcontinental railroad.[41]

In Santa Fe, Simpson was attached to the command of the military governor of New Mexico, Lt. Col. John M. Washington, and accompanied him on a punitive raid into Navajo country. Heading west out of Santa Fe, the expedition went to Chaco Canyon, Canyon de Chelly, south to Zuni and El Morro, and then back east to Albuquerque and Santa Fe. Simpson's report of the expedition included descriptions of the archeological finds and the first American account of El Morro. Simpson's assistants, Richard and Edward Kern and T.A.P. Champlin, provided illustrations, and Edward Kern drew the map which accompanied the report.[42]

In his report, Simpson recommended that an expedition be sent west from Zuni in search of a route to southern California which would be more direct than the two circuitous trails known at the time, the Emory-Cooke route to the south and the "Old Spanish Trail" to the north. The Old Spanish Trail, first "discovered" by William Wolfskill in 1830,[43] led northwest out of Santa Fe, crossed the Colorado River near present-day Moab, Utah, headed west to the Sevier River, followed the Sevier and the Virgin River southwest to the Las Vegas valley, then led on west across the Mojave Desert to Cajon pass and Los Angeles. It was used by Mexican traders, fur trappers, and many of the forty-niners.[44]

Simpson included in his report a description by trader Richard Campbell of an expedition from Santa Fe to Los Angeles and San Diego in 1827 along a trail across northern Arizona.[45] In September, 1851, Capt. Lorenzo Sitgreaves of the Topographic Engineers led an expedition west out of Zuni on a reconnaissance in search of the route described by Simpson and Campbell. He was accompanied by Lt. John G. Parke, also of the Topographic Engineers, S.W. Woodhouse as naturalist, Richard Kern as cartographer, Antoine Leroux as guide, and fifty infantrymen under Col. Edwin Sumner as an escort. The expedition started down the established trail along the Zuni River on September 24, 1851, then followed the Little Colorado River to the Grand Falls. Heeding Leroux's warning that following the Little Colorado any further would bring them to the floor of the Grand Canyon, Sitgreaves decided to leave the river and head due west, so as to strike the Colorado River below the Grand Canyon. They passed to the north of the San Francisco Mountains, to the

south of Bill Williams Mountain, on west across the plateaus, around
the north end of the Aquarius Mountains, through the pass between
the Cerbat and the Hualpai Mountains, and across the Black
Mountains, eventually reaching the Colorado River across from
where the California-Nevada border meets the river. Due to the poor
condition of the pack animals, the dwindling supplies, and the
injuries suffered from Indian attacks, Sitgreaves decided to head
south along the river to Ft. Yuma. After facing repeated attacks by
the Mojave and Yuma Indians, the expedition straggled in to Ft.
Yuma on November 30. After some rest and recuperation, they
continued their trek west to San Diego.[46]

Though Sitgreaves' expedition penetrated previously unexplored
territory, it was less significant than it might have been because it did
not establish the direct route from Zuni to Los Angeles across the
Mojave Desert as was its intent. Likewise, his report and map were
somewhat disappointing. The report lacked the usual scientific
generalizations on the geography and geology along the route traversed, and the map contained many misspellings and inaccuracies.[47]
On the bright side, however, Kern's drawings of landscape views
along the route and unique Indian tribes, such as the Yampais,
provided an important contribution to the knowledge of the west,
and Sitgreaves' map at least presented a more detailed picture of the
country between the Zuni and Colorado rivers than was previously
available.

Back in Washington, D.C. at this time, a many-sided, eight-year
deliberation over the location of a transcontinental railroad was
climaxing in Congress with a furious debate. It had been obvious to
all concerned that the economic condition of the country would
permit the construction of only one, or at the most, two transcontinental lines, and Congress had the task of deciding where the best
interests of the whole country lay. The complex pattern of rivalries
over securing the eastern rail termini, initiated by Asa Whitney's first
proposal for a transcontinental railroad in 1844, rendered the
legislature helpless in a political and economic deadlock, however. A
solution came in 1853 with a call for official surveys of the proposed
routes by Senator Richard Brodhead. The surveys, to be conducted
by the Topographic Engineers, promised to substitute the impartial

judgement of science for the passions of the promoters and politicians.[48]

One of these surveys was conducted from Ft. Smith to Los Angeles along the 35th parallel by Lt. Amiel W. Whipple. The party included Swiss geologist Dr. Jules Marcou (a protege of Louis Agassiz), artist-naturalist Heinrich Mollhausen (sent by Baron Humboldt), and Antone Leroux as guide. The primary objective of Whipple's survey was to examine the country between the Zuni villages and the Mojave River more carefully than had been done before, and to plot the exact route of the railroad.

The group assembled at Ft. Smith in July, 1853, after a long delay caused by the difficulty of obtaining surveying instruments. Many other groups had been recently outfitted, and Whipple had to wait in Washington while certain instruments were made. They finally left Ft. Smith on July 14th, and followed the Marcy-Simpson route along the ridge between the Canadian and Washita rivers, then on past Cerro Tucumcari, across the Pecos River at Anton Chico, and on to Albuquerque, where they were joined by an auxillary force led by Lt. Joseph C. Ives for protection through the hostile Indian country to the west.

The party left Albuquerque November 8 and surveyed various routes between there and the Zuni Pueblo. Heading on west, they explored and mapped a new route which joined the Rio Puerco and followed it to where it flows into the Little Colorado. After a few days march along the Little Colorado, Whipple decided to head west, hoping to find an easier route than Sitgreaves' by going south of the San Francisco Mountains, north of Bill Williams Mountain, and then down the Bill Williams Fork to the Colorado River. As it was, his trek proved to be as difficult as Sitgreaves', due to detouring around Canyon Diablo, near Meteor Crater in Arizona, and traveling in a circuitous path while trying to find the Bill Williams Fork.

After arriving at the Colorado River January 20, 1854, Whipple's expedition turned north and followed the river upstream to the Needles, crossing it there with the help of the Mojave Indians. Whipple divided his party into three for the Mojave desert crossing, due to the scarcity of water. The crossing was made without extreme

James A. Coombs

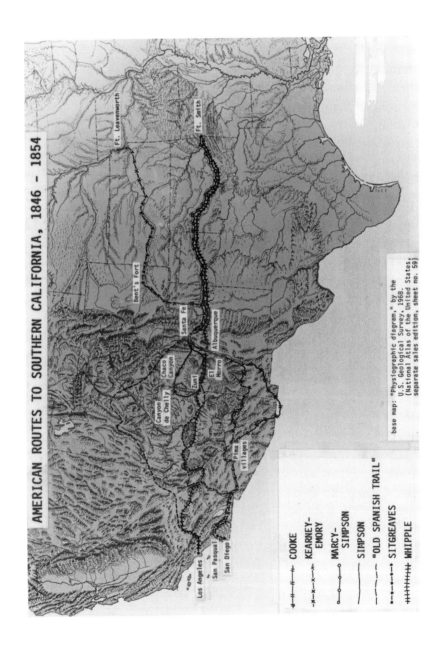

AMERICAN ROUTES TO SOUTHERN CALIFORNIA, 1846 – 1854

Ft. Leavenworth
Ft. Smith
Bent's Fort
Santa Fe
Albuquerque
Canyon de Chelly
Chaco Canyon
Zuni
El Morro
Pima villages
Los Angeles
San Pasqual
San Diego

COOKE
KEARNEY-EMORY
MARCY-SIMPSON
SIMPSON
"OLD SPANISH TRAIL"
SITGREAVES
WHIPPLE

base map: "Physiographic diagram" by the
U.S. Geological Survey, 1968.
(National Atlas of the United States,
separate sales edition, sheet no. 59)

hardship. They found the Mojave River about 25 miles southwest of Soda Lake and followed it until they joined the Mormon Road from Salt Lake City to San Bernadino, near present-day Barstow. This they followed through Cajon Pass and on to Los Angeles.[49]

After the party disbanded, Whipple took a boat to San Francisco, where he compiled his report. It was published in 1855, and included a large map in two parts. The first extended from the Mississippi River to the Rio Grande and the second extended from there to the Pacific Ocean. All of the information shown on the map, except for that in the vicinity of Whipple's route, was taken from previous Topographic Engineer maps, but it was still an important comprehensive map of the American West.[50]

In his report, Whipple was very enthusiastic, yet objective about the practicability of the 35th parallel railroad route. He described easily negotiable passes through the various mountain ranges and claimed the country was fertile and adaptable to agriculture, a much more optimistic view than previous explorers had reported. The impact of his report in Congress was diminished at first, however, because his cost estimate for building a railroad along the 35th parallel was much higher than that of the surveys of other routes. Whipple later revised his cost estimate so that it was more in line with the others, but proponents of the other routes used this to their advantage. As the 1850's progressed, though, more groups began to favor the 35th parallel route as a compromise solution to the location of the transcontinental railroad.[51]

DEVELOPMENT OF 35TH PARALLEL ROUTES TO SOUTHERN CALIFORNIA 1854-1926.

At this same time, military officers in Arizona and New Mexico, concerned with the slowness with which supplies reached their forts, petitioned the War Department for improvements. Consequently, Secretary of War John B. Floyd proposed to Congress a plan to move supplies by boat to the Colorado River and then overland on a wagon road across the 35th parallel to Ft. Defiance in New Mexico using camels as pack animals.[52] The concept of using camels in the American southwest apparently originated with George H. Crosman in the 1830's. He converted Army Quartermaster Henry C. Wayne, who in turn converted the previous Secretary of War, Jefferson Davis.[53] The motivation for using camels was related to the growing concern over the drawbacks of using horses, mules, and oxen to transport men and equipment on explorations and surveys. The use of these animals was actually threatening the success of the expeditions because of the threat of theft by Indians, the grueling punishment of climbing and descending rocky terrain, and death by starvation and thirst during the long desert treks. What the explorers needed was an animal that Indians would not covet, that could pack heavy loads, walk endless miles, feed on native plants of the desert, and survive for days without fresh water. The camel seemed to fit the bill.

In 1855, Congress approved and appropriated money to establish the Camel Corps,[54] and by August of the next year, the Corps was established at Camp Verde near San Antonio, Texas with 34 camels and five Arab caretakers acquired from northern Africa. After a year of acclimation to the Texas climate and people, the camels were ready for the task for which they were imported. Secretary of War Floyd's proposal to Congress was approved in the spring of 1857,[55] and Lt. Ed Beale was assigned to lead the survey of the wagon road, using camels as pack animals.

The surveying party assembled at Camp Verde, and departed June 19, 1857 with the 25 best camels from the herd, heading west to El Paso, then north to Albuquerque. It rained steadily for days, but the camels plodded along patiently while the men, horses, and mules were miserable, pulling wagons out of the mud. The camels subsisted on greasewood and other harsh shrubs, preferring them to grass, which demonstrated their ability to subsist easier than other pack animals in country where forage was scarce. Beale's only difficulty with the camels was that no one knew how to pack them properly, not even the Arab caretakers.

From Albuquerque, the expedition headed west along the 35th parallel, reaching Zuni on August 29. They had now reached the point where their real work would commence — the wagon road survey west to the Colorado River. With his railroad survey report and map in hand, Beale followed Lt. Whipple's route from Zuni to Leroux Spring at the base of the San Francisco Mountains, even following the faint tracks of his wagons at times. From there on west to the Colorado River, Beale blazed his own, more direct route, along which he found abundant grass, timber, and water. He also disproved the common belief that camels could not travel across rocky terrain, as they were unaffected by the rough volcanic rock of the north-central Arizona plateaus. In his diary, Beale predicted this road would become a great emigrant route to California.

The party arrived at the Colorado River about fifteen miles north of the Needles on October 17. Crossing it proved difficult, but less than Beale expected. He had been told that camels could not swim, and this seemed to be confirmed when the first camel led to the water's edge refused to get wet. Undismayed, Beale had the largest, healthiest camel brought up. It plunged right in and easily swam across. The other camels were then lined up, tied together in groups of five, and swum to the opposite shore without any difficulty. The other animals did not fare so well, as two horses and ten mules drowned in their attempt to cross the rapidly flowing river. Although the wagon road survey was completed at this point, the expedition continued west across the Mojave desert to Ft. Tejon, located in a mountain pass between the desert and the San Joaquin Valley, arriving there in early November.

On January 6 of the next year, Beale took twenty men and fourteen camels back to Ft. Defiance to test in winter the road he had just surveyed. They first went south to Los Angeles, then headed east through Cajon Pass and across the Mojave Desert to the Colorado River, where a rather bizarre accidental meeting occurred. Just a few minutes after they arrived at the river on January 23rd, a paddle-wheel steamboat appeared, piloted by George A. Johnson.

Johnson originated steamboat service on the Colorado between Ft. Yuma and the Gulf of California in 1853. When approached by the U.S. government in 1857 about opening navigation much farther up the river in order to supply troops being sent to Utah for the Mormon War, he demanded $3,500 a month for his services. Considering that excessive, the War Department built its own boat, the *Explorer,* in Philadelphia, then disassembled and shipped it around the Horn to the Gulf of California. There it was reassembled by an expedition led by Lt. Joseph C. Ives and used to survey the

Colorado up to the Grand Canyon.[56] Determined to be first up river, Johnson meanwhile steamed off on his own, where he met Beale and the camels. Beale was greatly moved by the chance meeting and remarked on the strangeness of the scene: camels standing on the riverbank in a wild and unknown country while a symbol of civilization, the steamboat, ferried Americans across the river.

After the steamboat incident the camels were returned to Ft. Tejon, but Beale continued eastward on his just-established road, which he found already in use by Indians. After fighting snowstorms and extremely cold weather, the party arrived at El Morro on February 19, and at Ft. Defiance soon after. Beale declared the camel experiment and road survey a success, enthusiastically reporting to Secretary of War Floyd how the capabilities of camels and their adaptation to the American southwest was proved through this practical demonstration. He was equally enthusiastic about the road, predicting that it would become the great emigrant route to California. Beale included the western half of Lt. Whipple's 1855 railroad survey map in his report,[57] to which he added his own route and that of Francois Aubry, who traveled from Tejon Pass to Albuquerque in 1853 and again in 1854.[58]

In 1858, Congress granted $75,000 to the War Department for the purpose of constructing bridges and improving the road from Ft. Smith to Albuquerque and on to California. Secretary Floyd determined the appropriation was insufficient to build a completed road across the route's entire length, so when he asked Lt. Beale to lead this road-building expedition, he instructed him to concentrate on improving the road's most difficult stretches.[59]

Beale's road party left Ft. Smith on October 28, 1858 on what one might expect to be a routine excursion up the Canadian River. About one hundred miles west of Ft. Smith, however, they met several mail stages waiting for them, enroute from Neosho, Missouri to Santa Fe. Recent fighting with the Comanche Indians had made it essential to travel with a military escort, and the stages were hoping Beale's party would provide one. Beale's own military escort had fallen behind, however, so while frigid winter weather encroached and provisions and tempers ran short, they all waited until November 26, when the escort finally showed up.

The party arrived at Hatch Ranch, just east of Albuquerque, on December 28 and stayed there to wait out the winter. Beale made trips to Albuquerque, Taos, and to Santa Fe, where he was amazed to learn that the Butterfield Overland Stage was running to El Paso, and on west to Los Angeles, along Col. Cooke's wagon road.[60] Before starting on the trek west again, Beale sent one of his men ahead to Los Angeles to acquire provisions to supply the party when it reached the Colorado River. The man found that the Mojaves living at the Colorado River crossing had become hostile, so while in Los Angeles, he asked for a military escort. When his request was denied, he appealed to S.A. Bishop, who was at Ft. Tejon using the camels for Army contract hauling.

Bishop gathered forty frontiersmen, loaded the supplies on camels, and headed for the Colorado. Met there by several hundred hostile Mojaves, they cached the supplies, crossed the river, and headed east to warn Beale. Beale was pleasantly surprised, to say the least, when Bishop rode into his camp a few miles west of the San Francisco Mountains on April 18 mounted on his favorite camel. On April 30, Beale's party arrived at the Colorado prepared for a fight. They were quite surprised to learn, however, that the Army had arrived ten days earlier and had signed a peace treaty with the Mojaves. The Army had also dug up Bishop's cache and stolen it, so Beale had to leave his work party at the river and go to Los Angeles to bring back another load of supplies. He arrived back at the Colorado on June 26, after spending some time at his ranch near Ft. Tejon, and began a return trip to Albuquerque with a road crew, which again improved the road along the way.[61]

On December 15, 1859, Beale submitted his final wagon road report to Secretary Floyd, and within a few years, his road became a major east-west commercial route. The Camel Corps program suffered a different fate, however. Despite Beale's and others' success of using camels as pack animals, the program lost direction and funding due to changes in command of the Corps and in the federal government's administration. The Department of War soon became almost totally occupied with the momentous issue of the southern states' secession. In the war that followed, the Army could not afford to experiment with camels. The Camel Corps was disbanded and the animals sold or turned loose in the desert. Various individuals and companies tried in vain to make a profit from the labor of camels, but the hostility of horses and mules, as well as that of teamsters and packers, frustrated their plans. The camels did not disappear quickly, though. Those camels turned loose in the desert provoked stories of wild camel sightings for more than three-quarters of a century.

Just as the national struggle over the expansion of slavery caused the demise of the Camel Corps, it also postponed any decision on the route of the transcontinental railroad, After the Civil War ended, however, many transcontinental railroad schemes were revived and a number of railroad companies were formed to link the midwest with California. One of these, the Atlantic and Pacific Railroad Company, received federal land grants in 1866 from Congress[62] to construct a line from Springfield, Missouri (the eastern terminus was later moved to Pacific, thirty miles west of St. Louis) to Albuquerque via the Canadian River valley, then along the 35th parallel across New Mexico and Arizona to Mojave, California, then across Tejon Pass to the San Joaquin valley and San Francisco. The line was to be completed by July 4, 1878, but progress was slow. By 1875, only 361 miles of track had been laid, from Pacific to Vinita in northeast Indian Territory, and the A & P had run out of money.[63]

A new railroad company, the St. Louis and San Francisco, emerged to revive the 35th parallel route. It built tracks from Pierce City, Missouri, fifty miles west of Springfield, to Wichita, Kansas, where it joined the Atchison, Topeka, and Santa Fe tracks which followed the old Santa Fe Trail through Raton Pass to Albuquerque. Through a complex financial arrangement, the Frisco and Atchison acquired the A & P right of way, and in the autumn of 1880, construction started on the line from Albuquerque to Needles along Beale's wagon road. 535 miles of track, as well as an iron bridge over Canyon Diablo, a tunnel in Johnson Canyon, west of Williams, Arizona, and a pile bridge across the Colorado River, were completed within three years. The Southern Pacific Railroad built tracks from Mojave, Californa to Needles in 1883 to connect with its line to San Francisco, and a new transcontinental line from that city to St. Louis was completed. By 1887, the Atchison had purchased some small Los Angeles area railroads and built tracks from Waterman (present-day Barstow), through Cajon Pass to San Bernadino, opening up a line from St. Louis to Los Angeles.[64]

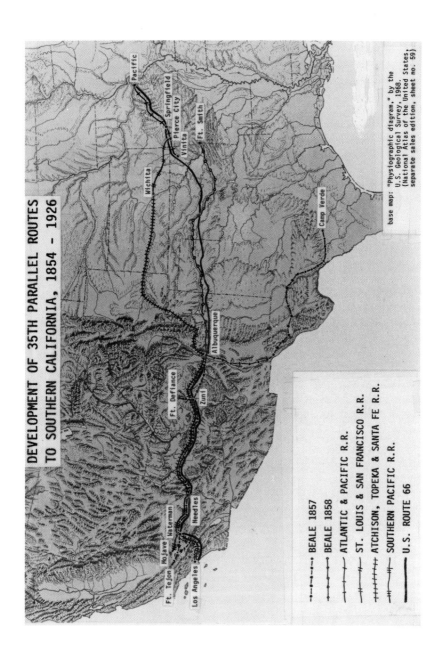

DEVELOPMENT OF 35TH PARALLEL ROUTES
TO SOUTHERN CALIFORNIA, 1854 – 1926

•—•—•—• BEALE 1857
•——•——• BEALE 1858
——+—— ATLANTIC & PACIFIC R.R.
—+++++— ST. LOUIS & SAN FRANCISCO R.R.
+++++++ ATCHISON, TOPEKA & SANTA FE R.R.
—#—#— SOUTHERN PACIFIC R.R.
——— U.S. ROUTE 66

base map: "Physiographic diagram," by the
U.S. Geological Survey, 1968.
(National Atlas of the United States,
separate sales edition, sheet no. 59)

CONCLUSION

The automobile became affordable to many Americans during the prosperous 1920's, and during the Great Depression of the thirties, many of them migrated west on a collection of roads which followed the 35th parallel to Los Angeles, officially designated as Route 66 in 1926. Businessmen in each state along the route formed Route 66 associations that petitioned state highway departments to pave the road, and by 1932, a hard surface road stretched from Chicago to Los Angeles, running through "Saint Louey and Joplin, Missouri . . . Oklahoma City . . . Amarillo; Gallup, New Mexico; Flagstaff, Arizona; . . . Winona, Kingman, Barstow, [and] San Bernadino."[65] The national Route 66 Association promoted the road as the quickest all-weather 2,200-mile route west from mid-America, and at one time, it claimed that 65% of all westbound traffic and 50% of all eastbound traffic used Route 66. Ironically, the Association's promotion eventually caused the Route's demise, and now the interstate highways which replaced Route 66 bypass the small-town businessmen who created the Association and who helped create the "Main Street of America."[66]

FOOTNOTES

1. Carl I. Wheat, *Mapping the Transmississippi West*, 5 vols. (San Francisco: Institute of Historical Cartography, 1957-1963), 1:29.

2. For a reproduction and detailed discussion of Martinez's map, see ibid, 1:28-33.

3. For a detailed discussion of Onate's 1604 explorations, see Harlan Hague; *The Road to California: the Search for a Southern Overland Route, 1540-1848*. (Glendale, Calif.: Arthur H. Clark, 1978), pp. 33-38.

4. U.S. Department of the Interior, National Park Service, *El Morro National Monument, New Mexico*. (Washington: U.S. Govt. Print. Off., 1965).

5. For a detailed discussion of Gracés' travels, see Hague, pp. 90-98. For a reproduction and discussion of his map of his travels, see Wheat, 1:92.

6. For a detailed discussion of the Domínguez-Escalante expedition, see Hague, pp. 98-103; Wheat, 1:94-99; and Ted J. Warner, ed., *The Domínguez-Escalante Journal,* trans. Fray Angelico Chavez. (Provo, Utah: Brigham Young University Press, 1976). For a detailed discussion and reproduction of Miera's maps, see Wheat, 1:99-116.

7. See Hague, pp. 105-120.

8. For a discussion and reproduction of this map, see Wheat, 1:117-119.

9. For more information on French expeditions to Santa Fe, see Noel M. Loomis and Abraham P. Nasatir, *Pedro Vial and the Roads to Santa Fe.* (Norman, Okla.: University of Oklahoma Press, 1967), pp. 28-73.

10. For more information on British penetrations into Louisiana and Spanish attempts to control its northern border, see Loomis, pp. 74-109.

11. Vial had explored routes between Santa Fe and San Antonio in the years 1786-89. For more information on these, see Wheat, 1:125-128 and Loomis, pp. 262-368.

12. Loomis, p. 371.

13. For a detailed discussion of Vial's expedition, see Loomis, pp. 369-407, and Hobart E. Stocking, *The Road to Santa Fe.* (New York: Hastings House, 1971), pp. 113-115.

14. For a reproduction and discussion of this map, see Wheat, 1:125-127.

15. For more information on these events, see Loomis, pp. xx-xxi.

16. For more information on Wilkinson's motives, see Donald Jackson, ed., *The Journals of Zebulon Montgomery Pike.* (Norman, Oklahoma: University of Oklahoma Press, 1966), 2:325n.-326n.

17. For a reproduction and discussion of this map, see Jackson, 1:458-459, plate 60, and Wheat, 2:11, 19-20.

18. Reproductions of these maps can be found in Jackson, between 1:324-325, 1:388-389, and 2:178-179; and in Wheat, 2:24-25.

19. For more information on Robinson's interests in Santa Fe, see Jackson, 1:376-379, 402-404; 2:192-193.

20. For more information on Walker and his maps, see Jackson, 1:413 and Wheat, 1:131 and 2:64-66.

21. Pike's journal, with annotations, appears in Jackson, 1:290-450.

22. Reproductions of these maps can be found in Jackson, 1:324-325, 1:388-389, and 2:178-179; and in Wheat, 2:24-25.

23. For detailed discussions of Pike's maps, see Jackson, 1:458-464; and Wheat, 2:20-27.

24. For a discussion and reproduction of Walker's map, see Wheat, 2:64-66.

25. For a discussion and reproduction of Robinson's map, see Wheat, 2:69-73.

26. Joshua Gregg, *Commerce of the Prairies: or the Journal of a Santa Fe Trader.* (New York: Henry G. Langley, 1844; reprint ed., Ann Arbor, Mich.: University Microfilms, 1966), 1:23.

27. For more information on Becknell and his adventures on the Santa Fe Trail, see Gregg, p. 21-24 and Archer Butler Hulbert, ed., *Southwest on the Turquoise Trail, the first Diaries on the Road to Santa Fe.* (Colorado Springs: The Stewart Commission of Colorado College; Denver: The Denver Public Library, 1933), pp. 56-68.

28. Use of Brown's manuscripts is acknowledged in William H. Goetzmann, *Army Exploration in the American West, 1803-1863.* (New Haven: Yale University Press, 1959), p. 110; in Gregg, pp. xxxi-xxxii; and in Wheat, 2:92-93n. Brown's field notes and General Sibley's diary of the survey are reproduced in Hulbert, pp. 107-174. Brown's maps are reproduced and discussed by Wheat, 2:87, 92-94.

29. Gregg, 2:155.

30. Besides the University Microfilms reprint cited above, *Commerce of the Prairies* has been reproduced as a Lakeside Classic, edited

by Milo Milton Quaife (Chicago: Lakeside Press, R.R. Donnelley & Sons, 1926) and in the American Exploration and Travel series, edited by Max L. Moorhead (Norman: University of Oklahoma Press, 1954).

31. Gregg's map is included in the reproductions of *Commerce of the Prairies*. It is also reproduced and discussed in Wheat, 2:181, 186-188.

32. Goetzmann, pp. 109-110.

33. For a more detailed account of Abert and Peck's expedition, see Goetzmann, pp. 123-127. For a reproduction and discussion of their map, see Wheat, 2:193-194.

34. For more information on the causes and political aspects of the War with Mexico, see Henry Ernest Haferkorn, *The War with Mexico, 1846-1848; a select bibliography* . . . (New York: Argonaut Press, 1965) or Ward McAfee and J. Cordell Robinson, *Origins of the Mexican War.* (Salisbury, N.C.: Documentary Publications, 1982).

35. Hague, p. 208.

36. The official report is: U.S. Congress. House. . . . *notes of a Military Reconnaissance from Ft. Leavenworth, in Missouri, to San Diego, in California,* by William H. Emory. H.R. Ex. Doc. 41, Serial 517, 30th Cong., 1st Sess., 1848.

37. This map, *Map of the Territory of New Mexico,* is discussed and reproduced in Wheat, 3:3,5. The original is published in Senate Ex. Doc. No. 23, Serial 506, 30th Congress, 1st Session, 1848.

38. For a more detailed account of Kearney's expedition, see Goetzmann, pp. 127-144; Hague, pp. 201-239; Emory's official report (see #36 above); and/or *Lieutenant Emory reports; a reprint of Lieutenant W.H. Emory's Notes of a military reconnaissance,* introduction and notes by Ross Calvin. [Albuquerque, University of New Mexico Press, 1968, c1951.]

39. This map, *Military reconnaissance of the Arkansas, Rio del Norte and Rio Gila,* is discussed in Goetzmann, p. 142; reproduced in sections in *Lieutenant Emory reports,* pp. 36, 72-73, 94-95, 112, 142, and 190; and discussed and reproduced in Wheat, 2:4, 6-7. The original was published in H.R. Ex. Doc. 41, Se-

rial 505, 30th Congress, 1st Session, 1848 and in Senate Ex. Doc. 7, Serial 505, 30th Congress, 1st Session, 1848.

40. For more information on Col. Cooke's Mormon Battalion and its wagon road, see Philip St. George Cooke, William Henry Chase Whiting, and Francois Xavier Aubry, *Exploring southwestern trails, 1846-1895.* (Glendale, Calif.: Arthur H. Clark, 1938), pp. 65-240; Philip St. George Cooke, *The Conquest of New Mexico and California, an historical and personal narrative.* (Albuquerque: Horn and Wallace, 1964); Hague, pp. 241-290; and Otis E. Young, *The West of Philip St. George Cooke.* (Glendale, Calif.: Arthur H. Clark, 1955), pp. 193-223. Cooke's official report, together with his simple map, was published with Emory's . . . *notes* in H.R. Ex. Doc. 41 (see #36 above).

41. For a more detailed account of Marcy and Simpson's expedition, see Goetzmann, pp. 213-218; Marcy's official report and map in House Ex. Doc. 45, Serial 577, 31st Congress, 1st Session, 1850; and Simpson's official report and map in Senate Ex. Doc. 12, Serial 554, 31st Congress, 1st Session. For a reproduction and discussion of their maps, see Wheat, 2:8,10-16.

42. For a more detailed account of Simpson's expedition, see Goetzmann, pp. 239-244 and Simpson's official report in Senate Ex. Doc. 64, Serial 562, 31st Congress, 1st Session, 1850 or published in James H. Simpson, *Journal of a military reconnaissance, from Santa Fe, New Mexico, to the Navajo Country . . .* (Philadelphia: Lippincott, Grambo and Co. . . . 1852). For a reproduction and discussion of his map, see Wheat, 2:9,16-17.

43. Goetzmann, p. 47.

44. for more information on the Old Spanish Trail, see Goetzmann, pp. 47-48; LeRoy R. Hafen and Ann W. Hafen, *Old Spanish Trail: Santa Fe to Los Angeles.* Vol. 1 of The Far West and the Rockies Hist. Ser. Glendale, Calif.: Arthur H. Clark, 1954); and Hague, pp. 139-145.

45. Goetzmann, p. 243 and Wheat, 2:20-21.

46. For a more detailed account of Sitgreaves' expedition, see Goetzmann, pp. 244-246 and Sitgreaves' official report in

Senate Ex. Doc. 59, Serial 668, 32nd Congress, 2nd Session, 1853.

47. This map, *Reconnaissance of the Zuni, Little Colorado, and Colorado Rivers . . . ,* is discussed and reproduced in Wheat, 3:23-24. The original was published with the report cited above.

48. For more information on the proposals for a transcontinental railroad, see John Debo Galloway, *The First Transcontinental Railroad.* (New York: Aron Press, 1981); Goetzmann, pp. 262-278; and Wheat, 3:176-203.

49. For a more detailed account of Whipple's expedition, see Goetzmann, pp. 287-289; Wheat, 4:17-20; and Whipple's official report in House Doc. No. 129, Serial 737-739, 33rd Congress, 1st Session, 1855, in Senate Ex. Doc. 78, Serial 768, 33rd Congress, 2nd Session, and in House Ex. Doc. 91, Serial 801, 33rd Congress, 2nd Session, 1856.

50. This map, *Explorations and Surveys for a Railroad Route from the Mississippi River to the Pacific Ocean,* is discussed and reproduced in Wheat, 4:77-80. The original is included with Whipple's report in the *Pacific Railroad Reports,* quarto edition, Volume XI, published as Senate Ex. Doc. 78 and House Ex. Doc. 91 (see previous footnote).

51. For more information on impact of the Pacific Railroad Reports and related developments in the location of the transcontinental railroad, see Galloway, pp. 41-51 and Goetzmann, pp. 295-304.

52. Odie B. Faulk, *The U.S. Camel Corps; an Army Experiment.* (New York: Oxford University Press, 1976), p. 97.

53. For more information on how these people were converted toward the concept of using camels in the southwest, see Faulk, pp. 18-20, 24-30.

54. Faulk, pp. 34-35.

55. Faulk, p. 97.

56. For more information on Lt. Ives' expedition with the *Explorer,* and the War Dept.'s dealings with Capt. Johnson, see Faulk, pp. 115-116; Goetzmann, pp. 378-394; Wheat, 4:95-98; and

Ives' *Report upon the Colorado River of the West, . . .* (Washington: Government Printing Office, 1861).

57. This map, *Preliminary map of the western portion of the reconnaissance and survey for the Pacific Rail Route near the 35 Par. made by Capt. A. W. Whipple T.E. in 1853-4. With additions, showing the route of the proposed wagon road from Ft. Defiance to the Colorado together with several lateral explorations,* is discussed and reproduced in Wheat, 4:94. The original is included with Beale's report in House Ex. Doc. 124, Serial 959, 35th Congress, 1st Session, p. 87.

58. For a more detailed account of Aubry's travels, see Philip St. George Cooke, *Exploring southwestern trails, 1846-1854.* (Glendale, Calif.: Arthur Clark, 1938).

59. Gerald Thompson, *Edward F. Beale and the American West.* (Albuquerque: University of New Mexico Press, 1983), p. 111.

60. Thompson, p. 115. For more information on the Butterfield Stagecoach Line, see Waterman L. Ormsby, *The Butterfield overland mail.* (San Marino, Calif.: The Huntington Library, 1942) or Frank A. Root, *The overland stage to California.* (Glorieta, N.M.: Rio Grande Press, [1970, c1901]).

61. For a more detailed account of Beale's second trip west, see Faulk, pp. 129-130 and Thompson, pp. 111-123.

62. Julius Grodinsky, *Transcontinental railway strategy, 1869-1893.* (Philadelphia: University of Pennsylvania Press, 1962), p. 9.

63. Keith L. Bryant, Jr., *History of the Atchison, Topeka and Santa Fe Railway.* (New York: Macmillan, 1974), p. 84.

64. For a more detailed account of the progress of construction on the 35th parallel rail route, see Bryant, pp. 84-105 and Grodinsky, pp. 166-169, 173-174, 195, 216-217, 259, 302.

65. Bobby Troup, "Route 66." Londontown Music, World Transcriptions, 1958.

66. "Get your kicks on Route 66," *Arizona Highways,* vol. 57, no. 7 (July 1981), pp. 5-34.

BIBLIOGRAPHY

Beiber, Ralph P., ed., *Southern Trails to California in 1849*. Glendale, Calif.: Arthur H. Clark, 1937.

Bryant, Keith L., Jr., *History of the Atchison, Topeka and Santa Fe Railway*. New York: Macmillan, 1974.

Cooke, Philip St. George, *The conquest of New Mexico and California, an historical and personal narrative*. Albuquerque: Horn and Wallace, 1964.

Cooke, Philip St. George; Whiting, William Henry Chase; and Aubry, Francois Xavier, *Exploring southwest trails, 1846-1854*. Glendale, Calif.: Arthur H. Clark, 1938.

Faulk, Odie B., *The U.S. Camel Corps: an Army experiment*. New York: Oxford University Press, 1976.

Galloway, John Debo, *The first transcontinental railroad*. New York: Aron Press, 1981.

"Get your kicks on Route 66," *Arizona Highways,* vol. 57, no. 7 (July 1981) 5-34.

Goetzmann, William H., *Army exploration in the American west 1803-1863*. New Haven: Yale University Press, 1959. maps.

Gregg, Joshua, *The commerce of the prairies: or the journal of a Santa Fe trader*. New York: H.G. Langley, 1844; reprint ed., Ann Arbor: University Microfilms, 1966.

Grodinsky, Julius, *Transcontinental railway strategy, 1869-1893*. Philadelphia: University of Pennsylvania Press, 1962.

Haferkorn, Henry Ernest, *The War with Mexico, 1846-1848; a select bibliography* . . . New York: Argonaut Press, 1965.

Hafen, Leroy R. and Hafen, Ann W., *Old Spanish Trail: Santa Fe to Los Angeles*. Vol. 1 of *The Far West and the Rockies Historical Series*. Glendale, Calif.: Arthur H. Clark, 1954.

Hague, Harlan, *The road to California: the search for a southern overland route, 1540-1848*. Glendale, Calif.: Arthur H. Clark, 1978.

Hulbert, Archer Butler, ed., *Southwest on the Turquoise Trail: the first diaries on the road to Santa Fe*. Colorado Springs: The Stewart

Commission of Colorado College; Denver: The Denver Public Library, 1933.

Jackson, Donald, ed., *The journals of Zebulon Montgomery Pike.* Norman, Okla.: University of Oklahoma Press, 1966.

Lieutenant Emory reports; a reprint of Lieutenant W. H. Emory's Notes of a military reconnaissance. Introduction and notes by Ross Calvin. [Albuquerque: University of New Mexico Press, 1968, c1951].

Loomis, Noel M. and Nasatir, Abraham P., *Pedro Vial and the roads to Sante Fe.* Norman, Okla.: University of Oklahoma Press, 1967.

McAfee, Ward and Robinson, J. Cordell, *Origins of the Mexican War.* Salisbury, N.C.: Documentary Publications, 1982.

Ormsby, Waterman L., *The Butterfield overland mail.* San Marino, Calif.: The Huntington Library, 1942.

Root, Frank A., *The overland stage to California.* Glorieta, N.M.: Rio Grande Press. [1970, c1901].

State Historical Society of Missouri, Columbia, MO. Abiel F. Leonard Manuscript Collection. Joseph C. Brown, "Chart and way bill from Tous to Fort Osage."

Stocking, Hobard E., *The road to Santa Fe.* New York: Hastings House, 1971.

Thompson, Gerald, *Edward F. Beale and the American West.* Albuquerque: University of New Mexico Press, 1983.

Troup, Bob, "Route 66," [song lyrics] Londontown Music, World Transcriptions, 1958.

U.S. Congress. House . . . *notes of a military reconnaissance from Ft. Leavenworth in Missouri, to San Diego, in California,* by William H. Emory. H.R. Ex. Doc. 41, Serial 505, 30th Cong., 1st Sess., 1848. maps.

U.S. Congress. House. *Reconnaissance and survey of a railway route from Mississippi River . . . to Pacific Ocean . . . ,* by Lieut. A. W. Whipple. Published in *Report of the Secretary of War on the several Pacific Railroad Explorations,* H.R. Doc. No. 129, Serial 737-739, 33rd Cong., 1st Sess., 1855. maps; and in Pacific Railroad

Reports, quarto edition, Vol. XI, H.R. Ex. Doc. 78, Serial 801, 33rd Cong., 2nd Sess., 1856. maps.

U.S. Congress. Senate. *Exploration of the Red River of Louisiana, in the year 1852*, by Capt. R.B. Marcy. Senate Ex. Doc. 64, Serial 562, 31st Cong., 1st Sess., 1853. maps.

U.S. Congress. Senate. *Report of an expedition down the Zuni and Colorado rivers*, by Bvt. Capt. L. Sitgreaves. Senate Ex. Doc. 59, 32nd Cong., 2nd Sess., 1853. maps.

U.S. Congress. Senate. *Report from the Secretary of War, communicating . . . the report and map of the route from Fort Smith, Arkansas, to Santa Fe, New Mexico*, by Lt. James H. Simpson. Senate Ex. Doc. 12, Serial 554, 31st Cong., 1st Sess., 1850. maps.

U.S. Department of the Interior, National Park Service, *El Morro National Monument, New Mexico*. Washington: Government Printing Office, 1965.

U.S. Department of War. *Report upon the Colorado River of the West, explored in 1857 and 1858 by Lt. Joseph C. Ives . . .* Washington: Government Printing Office, 1861.

Warner, Ted. J., ed., *The Dominguez-Escalante Journal*, trans. by Fray Angelico Chavez. Provo, Utah: Brigham Young University Press, 1976.

Wheat, Carl I., *Mapping the transmississippi west.* 5 vols. San Francisco: Institute of Historical Cartography, 1957-1963. maps.

Young, Otis E., *The west of Philip St. George Cooke.* Glendale, Calif.: Arthur C. Clark, 1955.

THE WAGON ROAD
SURVEYS

Charles A. Seavey

University of Wisconsin
Madison

R*eaders of this volume will*

quickly realize that the federal government was far more involved with the exploration and development of the trans-Mississippi West than a lot of popular, and some serious, history records. One of the reasons for this view goes back to the meeting of the American Historical Association in 1893 when a young Wisconsin historian arose to voice one of the enduring themes of American history.

> Stand at the Cumberland Gap and watch the procession of civilization, marching single file: the buffalo . . . the Indian, the fur trader and hunter, the cattle raiser, the pioneer farmer and the frontier has passed by. Stand at South Pass in the Rockies a century later and see the same procession . . . [1]

Frederick Jackson Turner was doing more than just propounding the frontier thesis of American history. He was listing the various groups who would be seriously studied by later historians. Soldier explorers and civilian government road builders did not go through the Cumberland Gap, but they certainly went through South Pass, as we shall see.

Beginning when President Thomas Jefferson dispatched Lewis and Clark on their trek, the government was intimately involved with western exploration, mapping and development. As the territory of the United States expanded westward and filled the continent, the government was not unaware of the vast potential wealth in the largely unexplored areas. Zebulon Pike, Stephen Long, John C. Fremont, William H. Emory, Joseph C. Ives, and John Wesley Powell were but a few of the numerous government employees who explored the west, and the maps they produced contain many cartographic milestones.

As the nation moved west, so did the government, encouraging exploration, and eventually, in 1838, creating a bureau to institutionalize that exploration. The Topographical Engineers, as part of the War Department, were the men primarily responsible for the "great reconnaissance."

While some government men got to do great and exciting things like explore and map the Grand Canyon, or see the sights and map Yellowstone National Park, others, largely unremembered today, divided the land up into one mile squares, or got involved in exploring for, mapping, and building wagon road surveys and maps that will be considered in this brief survey.

In the migration west, the plains of the midwest were not a major barrier. Except for minor items like Indians, lack of water in places, rather large rivers in others, and primitive means of transportation, the plains were little problem to the pioneers. Further west, the real problem began. No major waterway provided east to west transportation, and the transcontinental railroad was but a gleam in the eye of Thomas Hart Benton. Simply providing roads for wagons seemed the best alternative.

The "classic" period for road exploration and construction by the federal government was in the 1850's. There had been preliminary moves in Iowa, Arkansas, and other eastern states, and then a flurry of activity after the Civil War, but basically the '50's were the high water mark. There were two major federal agencies involved, and several geographic areas of activity.

The War Department, through its subunit the Topographical Engineers, mentioned above, was one of those agencies. The other was the Interior Department. Formed in 1849, that civilian agency immediately became involved in a prolonged struggle with the long established War Department over who was going to be responsible for the exploration of the trans-Mississippi West. The Interior Department eventually won this bureaucratic struggle with the creation of the U.S. Geological Survey in 1879, but it was not without great effort, and notable successes on the part of both agencies. At the height of the battle both departments operated subagencies known as the "Pacific Wagon Roads Office." This duplication lasted up to the Civil War, at which time both offices died a natural death.

Geographically there were a number of thrusts of wagon road exploration and construction. What might be called "internal" work involved surveying and construction in the Pacific Northwest, Minnesota, Utah, and west Texas. The Topographical Engineers were

mainly concerned with this internal system of roads. Goetzmann tells
us that the Engineers' roads were of:

> . . . predominantly local character. They were territorial roads,
> for the most part, and not continental projects. Only Simpson's
> trails across Utah had direct implications for long distance
> traveling[2].

Because of doubts as to the constitutionality of the federal
building of roads within states or territories, these roads were often
characterized as "military" roads. Although anybody could use them,
defense was, constitutionally, a federal duty[3].

The "overland," or long distance routes were mainly the work of
the Interior Department. Two major westward paths were involved.
The central Missouri Valley via South Pass to California, and the
southern Mississippi Valley via the Southwest, again terminating in
California.

Government involvement in exploration and construction of roads
did not start in the west. For example, in 1838, Lt. William Poole
surveyed 188 miles of "military road" from Saginaw to Mackinac,
Michigan. As a map, Poole's product is not much to look at (Fig. 14
Map list 1). The road is broken up into four segments running the
length of the printed page of the report, the southern terminus at the
top, north at the bottom, south to north running left to right on the
page. Each segment is provided with its own north arrow. The
segments, only about one centimeter wide, carry a running commen-
tary as to type of soil, terrain, and trees. Lt. Poole reckoned that the
cost of building the road through an apparently howling wilderness
to be $15,537! Although not terribly important in itself, the map is
indicative of federal government involvement in frontier
transportation.

At this time the Topographical Engineers were still firmly in
charge of exploration. Minnesota was an "experimental laboratory"[4]
where the Army built roads under the "military defense" guise in
the years before the Civil War. The Army interest in Minnesota
started before statehood. Captain John Pope led an expedition
through the then Territory in 1849. The result was a map at a scale
of 1:1,250,000 which shows all of Minnesota, and overlaps into

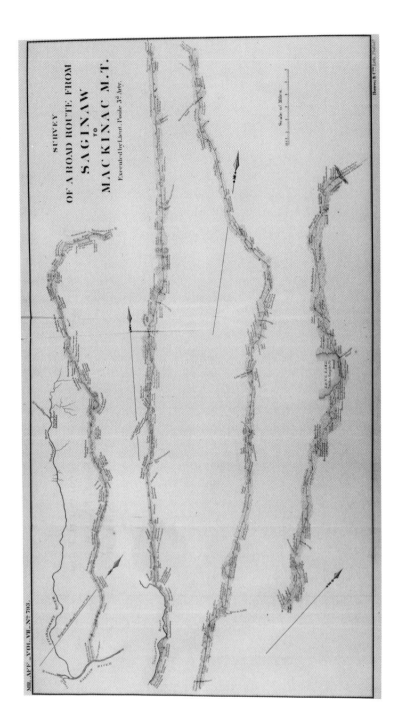

Fig. 14. "Survey of a Road Route from Saginaw to Mackinac, M.T."
Courtesy of the University of Kansas Government Documents and Map Library.

Iowa, Wisconsin, Canada, and South Dakota (Fig. 15 Map list 3). Aside from demonstrating the veracity of the "land of 10,000 lakes" tag for the state, the map shows federal concern with transportation. Portages and trails are noted, along with proposed lines of grants for railroads. Other items show concern with water transportation. The "Head of Navigation of Red River" is carefully marked, as are sand bar depths on Lake Superior at what is now Duluth. There are also little historical notes such as "where the enemy appeared on the hill," or "where the Kansas were killed." Captain Pope, the leader of the expedition became General Pope, lost the second battle of Manassas, and was fired by Lincoln. After the Civil War he became involved with wagon roads as an Army Department commander in the Northwest.

Once beyond Minnesota and across the plains the massive bulk of the Rocky Mountains tend to slow travel somewhat. A major breakthrough in westward movement was the discovery, by John Jacob Astor's "Astorian" expedition of 1810, of the great South Pass across the continental divide. South of the Wind River Range in southwestern Wyoming, South Pass was a conduit across the continent for some 350,000 people on their way west between 1841 and 1866. Today, South Pass is a strange and desolate area with little evidence of activity save the road itself, and a lone historical marker.

South Pass was, however, the focal point of a lot of exploration and mapping by the federal government. Federal interest in this central overland route lay in the fact that it was the most heavily traveled, and hence the most complained about, path to the west coast. From 1857 through 1860, the Interior Department mapped and worked on improving this segment of the road west.

There was a great deal of political maneuvering within Congress and the Interior Department as to who was going to be in charge of what. Eventually Albert H. Campbell was appointed "General Superintendent" of the newly created Pacific Wagon Road Office. The proposed road itself was divided up into several sections. The east and central divisions were assigned to William Magraw, to be assisted by F.W. Lander. The western division Superintendent was John Kirk.

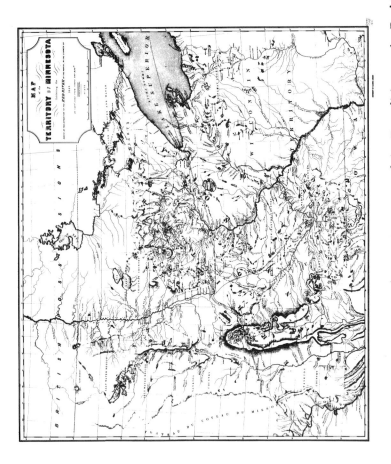

Fig. 15. "Map of the Territory of Minnesota Exhibiting the Route of the Expedition to the Red River of the North in the Summer of 1849."
Courtesy of the University of Illinois, Map and Geography Library.

On the east and central divisions, Lander was the principal explorer and worker in the years 1857-59. He was finally appointed Superintendent of those divisions in 1859.

There seems to have been more than the usual amount of political squabbling, accusations of mis-use of funds, and personal animosity among the various units at work. Things reached a low point in 1860 with a fist fight between Lander and William Magraw, whom he had replaced as superintendent of the project. A pistol was drawn, imprecations were hurled, but no permanent damage resulted, except to reputations. The road was improved, and water tanks built despite the political and personal problems.

The route under consideration went from Ft. Kearney (sic) on the Platte River in what is now Nebraska, through South Pass, and across to Honey Lake in what is now Lassen County, California. Honey Lake was the official end of the road in those days.

The maps, although nominally of the one road involved, are fairly detailed as to other trails, explorations, topography, and water sources. The two maps were prepared by different individuals, but are closely alike in style. The map of the "central division" covers the territory between South Pass and the "city of rocks" in modern Cassia County, Idaho (Map list 7). The map of the western division picks up at the city of rocks and takes the "New Worked Emigrant Road" to Honey Lake (Map list 6).

Further west the Army was involved in improvements of the routes within Utah. Although technically "internal" improvements the work in Utah did have implications for the overland travelers — a simple dictate of the geography of the situation.

In 1854, Congress authorized $25,000 for improving the road between Salt Lake City and San Diego, and the War Department commenced work. There were a number of problems with selection of routes, and finally in 1858 Captain James H. Simpson of the Topographical Engineers was directed to commence a survey of all routes within Utah.

The map showing routes explored and opened in Utah by Simpson is a reasonably typical product of the Topographical Engineers (Map list 5). Some attention is given to cultural features:

the grid for Salt Lake City is visible, for instance (five blocks by seven blocks!), as is that of Provo. Routes of earlier explorers are also shown (see below). The main interest is in wagon routes, however. The map legend lists the following routes in addition to those "Reconnoitred and Opened" by Simpson:

1. Wagon Route through Echo Canyon and over Big & Little Mountain.
2. Old Parley's Park Wagon Route.
3. Wagon Routes.
4. Capt. Stansbury's Route.
5. Capt. Beckwith's Route.
6. Also indicated are permanent and temporary camps.

Recommendations as to availability of water and other comments are also to be found on the map. "This route said . . . to be practicable for wagons after getting through Short Cut Pass" is one notation, for instance.

Eventually the War Department requested $170,000 for construction of roads within Utah: " . . . one of the most comprehensive programs of construction sponsored by the Topographical Engineers."[5] Congress, with the Civil War on its mind, never took action. The map, however, was already complete.

Simpson was a major figure among western explorers. He covered more of the west than any other member of the Topographical Engineers, including Fremont, and was solidly behind the wagon road program. His major report on the wagon roads was not published until 1875 when they had become passe, but it remains an excellent source of information. A Renaissance man in many ways, he laid some of the ground work for O.C. Marsh's later work in paleontology, yet he could also grasp and describe the geologic structure of the Great Basin[6].

What James Coombs refers to as the "Southwest Route" in another chapter of this volume, also figures in the wagon roads story. That route, and internal improvements in Texas were a major thrust of the wagon roads program. The Army was vitally concerned with communication in west Texas. Proximity to Mexico and quelling native Americans required rapid routes for military columns. The network in Texas was complex, and largely military in origin, although the Interior Department operated in this area too.

The first significant map of the area that we shall deal with was produced by William H. Emory, USTE, who would go on to survey the U.S.-Mexican border, and become a brilliant cavalry officer under Phillip Sheridan. On a map almost six feet long, Emory recorded General Kearney's march from Ft. Leavenworth, in what is now Kansas, along the Arkansas, down the Rio Grande, and across the Gila River to California during the Mexican War (Map list 2). Even under forced march conditions, Emory was noting possible wagon roads in his journal. These notes later appeared on his map. In one stretch south of the Gila River, for instance, Emory states: "Believed . . . to be an open Prairie and a good route if water can be had." There are frequent references to "rain water pools" or, less optimistically, "no water."

A later map of the eastern section of the same area by another Topographical Engineer suggests two alternate routes south of the Arkansas River (Map list 4). Randolph Marcy's map is not as large as Emory's, but shows the same attention to transportation possibilities. One route joins Ft. Smith, Arkansas and Santa Fe along the line of the Canadian River. The route then moved south along the Rio Grande and then west. The alternate route runs along the Red River, the Brazos, and the Pecos, south of the Llano Estacado, and terminates in Dona Ana, New Mexico, just north of present day Las Cruces. Both routes are heavily annotated with water locations. Springs and creeks are labeled, and ponds are either drawn in, or noted. There are even suggestions dealing with terrain that Marcy did not actually cover. North of the Llano Estacado we find this note: "Camanche Trail. Said to be good route for wagons, with water daily."

Under the same legislation that started work on the South Pass-Honey Lake route, the Interior Department commenced operations in the southwest in 1857. This project was accomplished without the drama that attended the construction of the northern road. James B. Leach was superintendent, with N.H. Hutton as his chief engineer.

Two maps, showing the El Paso, Texas to Fort Yuma, Arizona wagon road reflect this activity (Map list 8 and 9). The maps share stylistic similarities and concerns with the South Pass maps mentioned above. Although the stated western terminus is Yuma, map 9 depicts, in skeletal form, the route all the way to San Diego. These maps are nowhere near as elegant as Emory's depiction of the same area.

In the Pacific Northwest the Army worked in much the same fashion as in other "internal" situations. The construction of the "Mullan Road" from Fort Benton, Montana, to Walla Walla, Washington, was a major achievement of transportation and cartography. Lt. John Mullan planned and built well. His road is now largely covered by segments of Interstate 90! The maps were possibly the first by a government cartographer to make extensive use of contours, rather than the prevalent hachuring, to show relief. As such, Mullan and his maps deserve a bit of extended discussion.

Although the actual construction of the road was in 1859-1860, Mullan had been in the Pacific Northwest since 1853. On his first military assignment out of West Point he was attached to Issac Stevens' northern route explorations as part of the Pacific Railroad Surveys. Except for a tour of duty in Florida in the mid 1850's and brief trips to Washington, D.C., Mullan would remain in the northwest for the rest of his life.

The necessity for the Mullan Road actually arose because of steamboats. In the 1850's steamboats were ascending further up the Missouri River than ever before. At the same time, they were ascending the Columbia River to The Dalles, around 500 miles, as the crow flies, from the head of navigation on the Missouri. The military implications were obvious. If a road could connect the two river ports, then the Army would have much more rapid access to the Pacific Northwest.

After a false start in 1858, exploration and construction, fueled with a $100,000 Congressional appropriation, got under way in 1859[7]. Mullan and 230 men moved east from Walla Walla in July of that year. Along with the escort, engineers, and work crew, Mullan took Theodore Kolecki as topographer and map maker. Pushing to the northeast, the party both marked and improved the road as they went. By August 16th they reached Coeur d'Alene mission (now Coeur d'Alene, Idaho), having surveyed and improved two hunderd miles of road in six weeks.

The road east from Coeur d'Alene was not so easy, however. Thick standing timber and an "intricate tangle" of fallen trees slowed progress considerably. The path now had to be literally hacked out by crews of axmen working all day long. The 100 miles east of Coeur d'Alene took 15 weeks. Mullan finally gave up for the winter and established camp along the St. Regis Borgia River, east of present day Mullan, Idaho, on December 4th, 1859.

Work resumed, after the usual winter hardships, in mid-March, 1860. The road trended southeast to approximately present day Garrison, Montana, then turned northeast to present day Great Falls, and Fort Benton, Montana. The party arrived at Fort Benton on August 1, 1860.

Mullan was anxious to prove the utility of his road, and soon started the return journey, closely followed by a large party of Army recruits to be stationed in Washington. Mullan improved the road considerably on the westward trip. Crossings were graded, marshy portions were "corduroyed," or paved with logs, and some bridges were built. Following Mullan, the recruits marched into Ft. Walla Walla fifty-seven days and 600 miles after they left Fort Benton.

> Thus ended this military experiment via the upper Missouri and Columbia Rivers; and the success that attended it . . . the economy resulting . . . all constitute a sufficient commentary upon its feasability for future military movements toward the north Pacific[8].

Mullan's expectations for military usage were never really met. Jackson sums up the significance of the road this way:

> Although the Mullan road was ostensibly constructed for military purposes with the expectation that contingents and supplies for the Northwest posts could be moved with more rapidity and economy, it never attained importance as a through military highway . . . More important was its use as an emigrant route for those wishing to settle in the Northwest, or for the army

of mining prospectors who followed the gold rush into Idaho and Montana[9].

If the road was not important in the way it was expected to be, the maps were important under any circumstances. Three maps were published with Mullan's final report, issued in 1863. There are three sectional maps at 1:300,000, and an overall map at 1:1,000,000. The overall map was prepared by Edward Freyhold from Mullan's field notes (Fig. 16 Map list 13). It stretches from somewhat east of Fort Benton to Fort Dalles on the Columbia. Most detailed along Mullan's road, as might be expected, it contains a good deal of information on other routes as well. The "Colville Wagon Road," and "Capt. Frasier's Wagon Road" are among those noted. Typical of most Topographical Engineer maps of that time, it is drawn using hachures to show relief.

The three sectional maps, however, are not all typical. From west to east the maps, done at various times, cover portions of Mullan's road. The first dates from the abortive 1858 attempt to start the road (Map 10). It covers the area from Fort Dalles in the west, along the Columbia River valley to "Walla Walla City" and then follows "Captain Mullan's Military Wagon Road" north to the confluence of the Snake and the Palouse. The second map runs from that point, north and east to Coeur d'Alene (Map list 11). The third map picks up the trail at that point and follows it southeast to the northeast turn at Garrison, Montana, and thence to the Dearborn River, about sixty miles southwest of Great Falls (Map list 12). Not only do the sectional maps provide a detailed look at Mullan's road, but relief is shown, almost for the first time, in contours, rather than hachures. The western two maps have contour (or "curve") intervals of 100 feet, the third map an interval of 300 feet.

Wheat reports that Amiel Whipple had used contours on an inset in one of his Pacific Railroad survey maps, and that they had "occasionally" been employed as far back as 1822. But

> . . . never in the West, so far as we have noted, had an engineer shown the hardihood to adopt the contour system for portraying topography over a large area. [10]

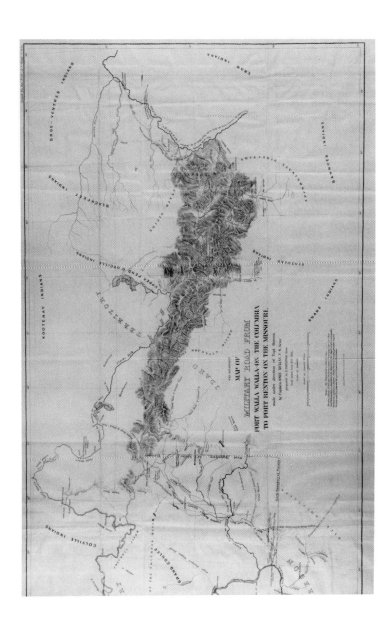

Fig. 16. "Map of Military Road from Fort Walla Walla on the Columbia to Fort Benton on the Missouri." Courtesy of the University of Kansas, Government Documents and Map Library.

Wheat reckons that the innovation was accomplished by Theodore Kolecki, the topographer employed by Mullan, who is mentioned in the title block of all three maps. Kolecki is directly credited with having "surveyed & drawn" the third map, and all three are very similar, stylistically, so Wheat is probably right. The decade of the 1850's (and a little beyond) closed on a high note, cartographically speaking.

As mentioned above, the impetus of wagon road exploration and construction really halted with the Civil War. The Topographical Engineers went on to their various fates, failure, glory or death with the Union or the Confederacy, and the Interior Department found its activities a distant second priority behind the prosecution of the war. Between 1850 and 1860, the Engineers, working on projects of predominantly local character, had constructed 34 separate roads costing a little over $1,000,000.[11] The Interior Department had mapped, and tried to improve the major overland routes to California, although incompetence at the managerial level did much to hamper their effectiveness.

After the Civil War there were attempts at cooperation between the War and Interior Departments, particularly in the Northwest. The long "overland" routes were mainly a thing of the past, and activity was not as intense as during the great decade of the 1850's. The race for Promontory Point was on, and the railroad had captured public and Congressional attention.

The federal effort in wagon road building had been fragmented, both geographically and administratively. It was not, however, without success, both in cartographic and transportation terms. We have considered some of the maps in this and other papers. Let John D. Unruh, historian of the westward migration, evaluate the effect on transportation:

> By 1860 overlanders (could travel) on trails which had been surveyed, shortened, graded, and improved by government employees. Overlanders even enjoyed the luxury of crossing bridged streams and watering their stock at large reservoirs . . . Especially in the 1850's federal presence in the west was far more important

to the success of the overland emigrations than (was then) acknowledged.[12]

Unruh's assessment is probably the best. After the wagon road explorations and constructions overlanders no longer had to winch their Conestogas up Windlass Hill. The old wagon tracks remain to this day, to be seen by tourists carried there on roads the likes of which Emory, Lander, or Mullan never dreamed.

This has been a painfully short study of a fairly complex subject. A full length study, W. Turrentine Jackson's *Wagon Roads West* (Berkeley, 1952) remains the definitive work on the explorations themselves. Although Jackson used the maps from the original reports as sources, he did not consider them in the text. Carl I. Wheat's *Mapping The Trans-Mississippi West, 1540-1861* (San Francisco, 1957-1963) remains the prime source for information on the maps themselves. He devotes a chapter, "The Wagon Road Program, 1857-1860" (vol. 4), to the South Pass Honey Lake road, and a substantial portion of another, "The Maps of 1863" (vol. 5), to John Mullan. With the exception of the Minnesota map, and Poole's survey in the *American State Papers,* all the other maps in this paper are mentioned by Wheat. The definitive history of the Topographical Engineers is William Goetzmann's *Army Exploration In The American West* (New Haven, 1959), and his *Exploration And Empire* (New York, 1966) is also very useful. John D. Unruh devotes a chapter to the federal presence in the west in *The Plains Across* (Urbana, III, 1979) the recent and definitive study of the transcontinental migrations.

Looming behind all of these authors is one of the greatest sources of all for American history. The U.S. Congressional Serial Set is a compendium of thousands of individual reports submitted to Congress since 1817. By 1900 it was over 3800 volumes long, and contained the reports of every explorer mentioned in this paper, and countless hundreds of others. Carto-historians cannot consider any aspect of 19th century American mapping without coming to the Serial Set.

NOTES

1. American Historical Association. *Annual Report*. Washington: Government Printing Office, 1893.

2. Goetzmann, William H. *Army Exploration in the American West, 1803-1863*. New Haven: Yale University Press, 1959. p. 350.

3. Jackson, W. Turrentine. *Wagon Roads West; A Study of Federal Road Surveys and Construction in the Trans-Mississippi West, 1846-1869*. Berkeley, Calf.: 1952. pp. 1-13.

4. Jackson, Chapter IV.

5. Jackson, p. 157.

6. Goetzmann, pp. 403-404.

7. Jackson, Chapter XVI.

8. Mullan's report, quoted in Jackson, p. 269.

9. Jackson, p. 273.

10. Wheat, Carl I. *Mapping the Trans-Mississippi West, 1540-1861*. 5 vols. San Francisco: The Institute of Historical Cartography, 1957-1963. Vol 5, part 1, p. 93.

11. Goetzmann, p. 347.

12. Unruh, John D. *The Plains Across; The Overland Emigrants and the Trans-Mississippi West, 1840-1860*. Urbana, IL: University of Illinois Press, 1979. pp. 201, 241.

LIST OF MAPS

1. *Survey of a road route from Saginaw to Mackinac, M.T. Executed by Lieutenant Poole, 3rd Art.* 25.5 × 49 cm. ca 1:170,000. Military Affairs vol. 7 of the *American State Papers*, Washington: Gales and Seaton, 1832-1861. This map is listed as ASP 25 in my "Maps of the American State Papers", Special Libraries Association, Geography and Map Division *Bulletin*, no. 107, March, 1977.

2. *Military Reconnaissance of the Arkansas, Rio del Norte, and Rio Gila, by W.H. Emory . . . 1847.* ca. 1:1,500,000. 80 × 178 cm. In: 30th Congress, 1st Session, Senate Executive Document 7, or

House Executive Document 41, Serial Set volumes 505, 517. Wheat no. 544.

3. *Map of the Territory of Minnesota Exhibiting the Route of the Expedition to the Red River of the North in the summer of 1849, by Captain John Pope, Corps Topl. Engrs. ca.* 1:1,250,000. 65 × 74 cm. In: 31st Congress, 1st Session, Senate Executive Document 42, Serial Set volume 558. Not listed in Wheat, but is listed on page 432 of Philip Lee Phillips' *List of Maps of America,* Washington, GPO, 1901.

4. *Topographical Map of the Road from Fort Smith, Arks., to Santa Fe, N.M., and from Dona Ana, N.M. to Fort Smith . . . by Capt. R.B. Marcy. ca.* 1:2,300,000. 42 × 77 cm. In: 31st Congress, 1st Session, House Executive Document 45, Serial Set volume 577. Wheat no. 681.

5. *Preliminary map of Routes Reconnoitred and opened in the Territory of Utah by Captain J.H. Simpson . . . 1858. ca.* 1:322,000 77 × 113 cm. In: 35th Congress, 2nd Session, Senate Executive Document 40, Serial Set volume 984. Wheat no. 957.

6. *Dept. of the Interior Pacific Wagon Roads Map of the Western Division of the Fort Kearney, South Pass and Honey Lake Road, surveyed by F.A. Bishop . . . 1857.* 1:720,000. No measurement attempted because of the extremely poor condition of the map. In: 35th Congress, 2nd Session, Senate Executive Document 36, Serial Set Volume 984. Wheat no. 966.

7. *Dept. of the Interior Pacific Wagon Roads Preliminary Map of the Central Division, Fort Kearney, South Pass & Honey Lake Wagon Road. Surveyed . . . under the direction of F.W. Lander, 1857-1858. ca.* 1:720,000. No measurement possible. In: same as no. 6. Wheat no. 1004.

8. *Dept. of the Interior Pacific Wagon Roads Map no. 1 of the El Paso & Ft. Yuma Wagon Road. Made under the direction of N.H. Hutton . . . 1857-8.* 1:600,000. No measurement possible. In: same as no. 6. Wheat no. 981.

9. *Dept. of the Interior Pacific Wagon Roads Map no. 2 of the El Paso & Ft. Yuma Wagon Road.* The rest of the data are the same as no. 6. Wheat no. 982.

10. *Map of Military Reconnaissance from Fort Dalles, Oregon, via Fort Walla Walla to Fort Taylor, Washington Territory . . . by Lieut. John Mullan . . . 1858.* 1:300,000, 61 x 93 cm. In: 37th Congress, 3rd Session, Senate Executive Document 43, Serial Set volume 1149. Wheat no. 1077.

11. *Map of Military Reconnaissance from Fort Taylor to the Coeur d'Alene Mission, Washington Territory . . . by Lieut. John Mullan . . . 1858.* 1:300,000. 63 x 55 cm. In: same as no. 10. Wheat. 1078.

12. *Map of the Mountain Section of the Ft. Walla Walla & Fort Benton Military Wagon Road from Coeur d'Alene Lake to the Dearborn River, Washington Territory . . . by Capt. John Mullan . . . 1859-1863.* 1:300,000, 60 x 123 cm. In: same as no. 10. Wheat 1079.

13. *Map of Military Road from Fort Walla Walla on the Columbia to Fort Benton on the Missouri . . . by Captain John Mullan . . . 1858-1863.* 1:1,000,000. 50 x 90 cm. In: same as no. 10. Wheat 1080.

GEORGE M. WHEELER AND THE GEOGRAPHICAL SURVEYS WEST OF THE 100TH MERIDIAN 1869-1879

Robert W. Karrow, Jr.

The Newberry Library
Chicago

In the decade before the United

States Geological Survey was established in 1879, four governmental organizations, sometimes in competition, each with its own special flavor and areas of expertise, were engaged in activities that included mapping the American west. The "four great surveys" were actually preceded by another government exploration and mapping effort that, in area covered, scope of investigation, creative mixture of government and civilian personnel, and in the contents and sheer bulk of its publications, deserves to be ranked as their progenitor. This was the Pacific Railroad Survey of 1853-1855, which set many of the patterns that later surveys were to follow and the crowning cartographic achievement of which, G. K. Warren's map, has been called "the culmination of six decades of effort to comprehend the outlines of western geography."[1]

After the cataclysm of the Civil War, and with railroads building from Sacramento and Omaha toward their junction on the shores of Great Salt Lake (though not, incidentally, following any of the routes explored by the Pacific Railroad Survey) the broad locational questions about the west had been answered. The new surveys, all launched by 1870, had different goals. The Geological and Geographical Survey of the Territories, brainchild of geologist Ferdinand Vandeveer Hayden, had begun work on the northern plains in 1867. Hayden was scientifically respected and enjoyed a great deal of popular support among influential westerners. The appropriations for his survey, under the auspices of the Interior Department, were large for the time, and, with a total cost of some $700,000 for the twelve years of the survey, Hayden's ranks as the most expensive of the efforts.[2] Most of his labors went into studying the geology, botany, ornithology, and entomology of the west, and his maps were all intended to illustrate geological structures. The primary cartographic contribution of the survey was its *Atlas of Colorado*, published in 1877.[3]

A second survey, comparable with Hayden's in that its prime mover was a geologist and its prime motivation mineral exploration, was the Geological Exploration of the Fortieth Parallel, begun in 1869 under the direction of Clarence King. King was a *wunderkind*, heading a major government exploration at the age of twenty-seven, a product of Yale's Sheffield Scientific School, apprenticed for four years to the influential J.D. Whitney's California Geological Survey, hailed as one of the best and brightest men of his generation, and later the first director of the newly formed U.S. Geological Survey. King's survey was a practical exercise in economic geology, aiming to determine the economic potential of the route followed by the transcontinental railroad through the Great Basin. Its chief cartographic product was an atlas of eleven maps.[4] His survey, under Army auspices, but with exclusively civilian personnel, was the only one of the four not legislated out of existence in 1879; it had closed its books the previous year.[5]

The third survey was the work of the only man of the four directors whose name is still commonly known, John Wesley Powell. This one-armed geologist, ethnologist, writer, and teacher first sprang into the public attention with his 1869 boat trip through the Grand Canyon, probably the last great exploit in exploring the unknown in sub-arctic North America. As the second director of the U.S. Geological Survey, as head of the Bureau of American Ethnology, he followed his chief loves, geology and native peoples, and left a legacy of reports, collections, and students whose influence continues today. Almost more importantly, Powell was a skilled administrator, no less at home in a Washington office or in the halls of Congress than in an Indian pueblo or in the "Canyon of Lodore." His survey, an outgrowth of the 1869 river trip, was called the Geographical and Geological Survey of the Rocky Mountain Region, although its actual area of operations was primarily the Colorado and Green River valleys and the Great Plateau of Utah. The mapping done by the Powell survey was, after an inauspicious beginning, careful and with a relatively high level of control. Most of the maps, however, remained in manuscript when the survey was officially disbanded and ended up in the offices of the Geological Survey, where they continued to be used as sources for many years.[6] The published record of the Powell survey stressed geology and ethnology. One of its publications, Powell's monograph on the *Arid Regions of the United States,* was a seminal work on water rights and land subdivision in the west.

Fig. 17. Lt. George Montague Wheeler, ca. 1870.
Courtesy of the Henry L. Abbot Collection, U.S. Army
Engineer Museum, Ft. Belvoir, Va.

It was with this background, and in this company, that the Army
entered the field of western mapping with the Geographic Surveys
West of the One Hundredth Meridian. The surveys were instigated
by a young first lieutenant, George Montague Wheeler, who found
himself, in 1869, assigned to survey some 24,000 square miles of
Nevada desert. He produced a printed map[7] and in the following
year conceived a way to continue his work in the west and to make a
unique contribution to American cartography. He would map the
entire west, but not haphazardly, in support of geological or other
researches; he would map the entire region systematically, piece by
piece. He proposed to concentrate on the area west of the one
hundredth meridian (as good a natural definition for the beginning
of the "west" as any, then or now) and divided it into 95 rectangles,
each covering 2° 45′ of longitude and 1° 40′ of latitude. He
proposed to map this area of 1,400,000 square miles at a uniform
scale and to produce what we would today call general-purpose
topographic maps; maps that would show, not the geology, but the
shapes and arrangement of the landforms and river systems, and the
human constructions — roads, railroads, telegraph lines, cities,
towns, mines — that connected and punctuated them. These maps
would be intended primarily for the military, but would be available
to all and would serve the needs of most citizens for immigration,
business and industry, and agriculture.

There was nothing new about the general idea: France became the
first nation to complete its topographic map, in 182 sheets, in 1760
and by the 1870s most countries in Europe had similar topographic
coverage. Furthermore, all these European productions were the work
of the military branches of government. Wheeler was simply propos-
ing to Americanize a well-established cartographic genre, but alone
among his contemporaries, he took the big step of drawing the sheet
lines on the map, lobbying for the funds, planning the work, and
beginning its execution.[8]

Wheeler must have wanted badly to be an Army officer. Although
born in Massachusetts and apparently resident there, he went to live
with a brother in Colorado, hoping, one suspects, that nomination to
the military academy would be more certain from that relatively
unpopulated territory. He was the first appointee to West Point from

Colorado, and spent the Civil War there, graduating in 1866 with distinctions: he ranked first in philosophy and mathematics, second in engineering, and sixth overall.[9] But the young officers in Wheeler's class had just missed participating in the supreme military contest and their careers would be affected and often retarded as they competed with older veterans, many of them volunteers with brevet commissions, for places in the peacetime army. In addition, after the Civil War, the army entered a period of decline that was to persist almost until the end of the century. Reacting to the agony of war, an anti-military atmosphere dominated national thinking. The army was slashed in size and its personnel depressed by low pay, poor living conditions (especially in the West) and a policy of promotion that resulted in advancement at a glacial pace. It is not surprising that the quality of enlisted men was low or that one-third of the enlisted men deserted.[10] Some of these tensions are illustrated in an event that took place in one of Wheeler's field parties in 1872. The officer in charge, a lieutenant named Hoxie, had warned that he would "douse" any man still asleep after Hoxie had risen. He used this technique once, with a soldier named Harris:

> Emptied one cup of water over his face. He got up and assailed me with mutinous and abusive language — "Now see here, Mr. Hoxie, that's played out. By God, if you ever do that again it will be worth your commission to you. By God, you'll never get back to Washington. Treat a man like a dog. Come and pour water on him when he's asleep." I intimated my unconcern at his threats and ordered him to shut up his mouth and get up. He replied, "I won't shut up for you or any other man. I'll talk just as long as I damn please." And he continued in this same strain until he was dressed and had walked over to the fire. I then ordered him to go and help bring in the mules. He replied "I'll go when my shirt gets dry," but eventually went off to obey the order, muttering threats on the way. "By God, you wouldn't do this in civil life. You'd get your God-damned head put [off] you." Before this I told him he was saying too much and could be convicted now of mutiny. "I don't give a good God damn. — This world is wide enough to get around in" & intimated that it was an easy matter

to escape the consequences of what he was doing by deserting . . . ."[11]

The tension blew over and a few weeks later Hoxie issued Harris a pistol.

Another of Wheeler's lieutenants, operating near Albuquerque in early November, found himself one morning with a real enough mutiny, this time of the civilian packers. They were anxious to reach Santa Fe, where their duties would end for the season, and were convinced that the young lieutenant was going the wrong way. They threatened to quit then and there unless he changed his course. He wrote: "There were eight of these packers, all frontiersmen. My enlisted men I think, were as worthless a lot as ever served the Government, but they all felt a deep antipathy to the packers."[12] He was able to capitalize on this manifestation of the pecking order, and got the packers to agree to stay on for another day or two. By early afternoon, they struck the main road and the mutinous talk disappeared. Life in the frontier army was not easy and tempers must have been frequently short. One incident of undue reaction under pressure was to surface in the press and cause embarrassment to the survey, but on the whole, the record is free from overt conflicts, and it is surprising that so much good work was done under such trying circumstances.

Wheeler launched his second western season (and the first with his newly-created survey) in 1871. His field of operation was to be the Southern California desert and adjacent areas of western Arizona. Some 72,250 square miles of the most forbidding terrain in North America, including Death Valley, were surveyed that summer, but Wheeler and a small party commenced, late in the year, on what, considering the overall achievements of the survey, can only be considered an ill-advised side trip. Leaving Camp Mojave, Arizona, on the lower Colorado River in four boats, the party proceeded to row, pole, push and drag the boats for two hundred miles against the current. Their destination was the mouth of Diamond Creek, well inside the lower Granite Gorge of Grand Canyon, and, after a month of the most grueling work, the last week on short rations due to the loss of one of the boats, the party reached their goal. The return trip

Fig. 18. In camp near Belmont, Nevada, 1871.
(National Archives 106-WB-21)

took just five days. Wheeler may have hoped that the expedition would place his name in the ranks of the great explorers of the American West, but the trip was one that did not need to be made. Powell had descended the river through the whole length of the canyon in 1869 and was coming through the upper canyon again in the same month as Wheeler's trip. The part of the river he ascended could not be called unknown. In his final report (written in 1879) Wheeler claimed as one of the reasons for the trip the determination of the "absolute limit of navigation" but this site (near present Pierce Ferry, Arizona) was reached by the seventeenth day of the trip. The rest was an exercise in sheer willpower, motivated by the kind of desire that fuels mountain climbers. Wheeler concluded in his report that "the exploration of the Colorado River may now be considered complete,"[13] but later historians have not been as willing to grant importance to the expedition. Richard Bartlett called it foolish, and G.K. Gilbert's biographer has written: "Where Powell leaped to public prominence by traversing the dramatic canyons of the Colorado River, Wheeler reversed the procedure. But travelling upstream reversed the consequences as well, yielding consternation and ridicule rather than glory. Here, indeed, is the symbol of the Wheeler survey: conceptually and institutionally, no less than geographically, it traveled against the mainstream."[14] At least this artificially created adventure, this attempt by Wheeler at the grand exploratory gesture, came early in his career. With his desire for exploration sated by the river trip, Wheeler felt free to devote the remaining years of his survey to the more prosaic but infinitely more useful and monumental tasks of mapping the ground. The originality of Wheeler's work lay in his conception of a uniform and systematic map of the West, and it was precisely in "traveling against the mainstream," in (after the River trip) eschewing exploring expeditions and in downgrading the geological and other aspects of his survey, that he made his greatest contributions. In its geodetic and mapping work, the Wheeler survey was well beyond ridicule.

The principles underlying Wheeler's surveys had been known and practiced for centuries and were basically unchanged until the advent of aerial photography in the 1920s, although each generation brought refinements in instrumentation and subsequent improvement in accuracy. The first step in the topographic survey of a large area is the establishment of several accurate positions. Once a given set of points on the earth's surface is accurately determined, triangulation and compass traverses can be used to fill in the areas between these points. Positions are determined by reference to the imaginary grid of lines of latitude and longitude which may be imagined to lie on the surface of the earth. Any given point can be defined as so many degrees, minutes, and seconds north or south of the equator, and so many degrees, minutes, and seconds east or west of the Prime Meridian running through Greenwich, England. Because the equator and the poles are natural features, defined by the axis of the earth's rotation, the determination of latitude (degrees north or south of the equator) is relatively straightforward and has been accomplished with ease for hundreds of years. In the simplest case, in the Northern Hemisphere, an observer measures the distance between the horizon and Polaris, the star nearly above the north pole. In the Wheeler survey, the latitude of primary and secondary astronomical stations was established by averaging many sightings of many pairs of stars. At the station in Pioche, Nevada, in 1872, the three astronomers on duty made 193 observations of dozens of stars over a period of seven nights. They arrived at a latitude of 37° 55′ 26.07″ (± 0.05″). Their probable error amounts to ± five feet on the ground.[15]

The determination of longitude, the distance east or west of the Prime Meridian running through Greenwich is much more problematical, although the basic principle is readily explained. If we wish to know how far west we are from the Greenwich meridian, we must determine the difference in time between our location and Greenwich. For instance, four hours is one-sixth of a day, so four hours' difference between the two locations would mean a difference of one-sixth of the circumference of the earth, or 60° of longitude, between them. The principle is simple, but as a practical matter, the determination of longitude with sufficient accuracy for use in navigation and surveying proved to be one of the major scientific hurdles of the modern era. The difficulty is that the speed of rotation of the earth is such that very accurate timepieces are needed to obtain a reasonably accurate reading of longitude. If we imagine that we are standing above the earth on a fixed platform at 39° north latitude and looking down at the earth moving beneath us, one degree of longitude, or about 54 miles, would pass beneath us in only four minutes. One second would see almost a quarter of a mile passing beneath us; one quarter-second, about three hundred feet. Obviously, to be able to determine longitudinal positions to within a few feet, as was possible with latitude, extremely accurate determinations of local and Greenwich time would be necessary.

In the Wheeler survey, time signals were received over the telegraph from Salt Lake City, connected with the Lake Survey observatory in Detroit, with the Naval Observatory in Washington, and, ultimately, via the transatlantic cable, with the observatory in Greenwich. In the most elaborate set-ups, such as at the Ogden, Utah station in 1873-74, a chronograph was used for recording time signals. The chronograph has been described as a device for "measuring time by the yard": it consisted of an electrically operated pen which translated the pulses coming over the telegraph line into marks on a rotating drum.[16] Local time was computed, as was latitude, by averaging many celestial observations. Survey personnel even invented a device for calculating the "personal equation," that is, the personal response time between the moment the observer sees a star pass the cross-hairs of his telescope and the moment he presses his telegraph key to record the event on his chronometer. Using such careful methods, and including elaborate corrections for instrument errors, the Wheeler survey was able to calculate longitudes at primary stations with an average probable error of $\pm\frac{1}{3}$ second.[17] This translates to ±30 feet on the ground. Such accuracy was possible only at the main stations, where there were permanent observatories, with masonry pillars for mounting the transit and optimum working conditions. But observations for latitude and approximate longitude were also made at secondary stations established at convenient points in the field, usually on mountain peaks. One such scene is described in a contemporary account:

> . . . when the other members of the party are chatting around
> the camp fire, the astronomer retires to a dark, quiet spot with his
> instrument, and lying on his back, or resting his elbow on the
> rough ground, occupies himself with the stars, which never seem
> so cold, so far, nor so brilliant elsewhere as they do from a peak in

Fig. 19. Survey observatory at Ogden, Utah, 1871. (National Archives 106-WB-294)

the Rocky Mountains or in the Sierra Nevada. He is assisted by an observer for time, who, with an open watch and a lantern before him, records the hour, minutes and seconds in response to the word "tick," which the astronomer utters at each observation. These two men, isolated and scarcely revealed by the flash of the fire and the yellow gleam of their own lanterns, make a picture, and when the night is frosty, the picture is one of misery.[18]

There were many means of discomfort available to the field parties, one of which included carrying transits and tripods weighing upwards of forty pounds up 10,000-foot mountains. On one occasion, a party found itself on a terrace with no way to go but down a long, snowy incline. They were planning to cut steps in the snow when the astronomer, we are told;

> . . . incautiously stammped his heels on the edge to try its brittleness. His foot slipped from under him, and the next moment we were thrilled by seeing him sliding down the mountain with the velocity of a flash of light. He was in a sitting posture, his hair was blown back, and his hat slowly rolled down after him . . . he carried a spiked tripod, which made an excellent alpenstock, and with fine presence of mind he plunged this into the snow between his legs, slid half way up it, and suddenly came to a stop.[19]

The astronomical work of Wheeler's survey was of a very high order for its time and place, and John Wesley Powell, though he later did his best to see the survey disbanded, was honest enough to remark that "[Wheeler's] astronomic work ranks with the best that has ever been done in this country, and, perhaps, with the best that has ever been done in the world."[20] A star catalog published by the survey in 1879 became a standard reference work, still in use twenty years later by the U.S. Geological Survey.[21]

With a few widely-spaced points fixed very accurately, other prominent features can be related to these by triangulation. Triangulation, first used in surveying by Gemma Frisius in the sixteenth century, is based on the principle that if one knows the length of one side of a triangle and the angles at either end, the lengths of the other two sides can be determined. Triangulation is normally done between mountain peaks, and the idea is to create a network of "strong" triangles, as close to equilateral as possible, covering the entire area to be surveyed. The first step is the measurement of one side of a triangle. The line so measured is called the base line, and extreme accuracy is required because upon it will depend the accuracy of all subsequent angular measurements. The base line is laid out on a level surface, and the Wheeler survey used a specially constructed wooden rod, twenty feet in length, for measuring it. The ends of the rod were raised off the ground on steel plates in each of which was inset a silver plate with a hairline ruled in it. Eight inches at each end of the rod were graduated into 100ths of an inch, and an observer at each end of the rod read the point at which the graduations crossed the hairline ruled in the plate. With a magnifier, they could read to 1/1,000 of an inch. The survey used wooden rods because they were less subject to expansion and contraction than steel rods or tapes; still, they made regular thermometer readings during measurement to correct for the expansion of the rod, calculated to be 9/10,000 of an inch for each degree Fahrenheit. Repeating this process 1,181 times on a base line near Sutro, Nevada, they measured a four and one-half mile line in twelve days. Then they turned around and measured the same line in the opposite direction, finishing in eight days. The difference between the two measurements was a little more than 3/100 of an inch.[22] From such base lines, the network of triangles was developed, with sides measuring twenty to seventy miles. Each point was occupied in turn and the system of triangles gradually extended.

From each peak the topographer made a profile of the entire horizon to locate other peaks used in triangulation and to serve as a guide in drawing the finished map. Various secondary positions were established within the large triangles by occupying the points or by sighting to them from three other stations. A horizontal sketch, using contours, was also made at each triangulation station, paying particular attention to the drainage. These contours, being sketched in and not based on absolute measured elevations, are more properly described as form lines, a distinction which Wheeler was quick to make.

Within the framework created by the triangulation, detailed topography was filled in by the meander method. In this technique, a

THE ODOMETER CARRIAGE.

Fig. 20. "The Odometer Carriage," from *Harper's New Monthly Magazine* 55 (1877): 68. Courtesy of The Newberry Library.

party proceeded along roads or trails, measuring angles at turns of the road and to nearby peaks with a small transit, and sketching and making notes on the surrounding topography. Meanwhile the distance along the road was being recorded with an odometer, a large wheel of known circumference attached to a counter. The Wheeler survey brought the meander method to a high degree of reliability; still, the odometer, pulled by an army mule, not uncommonly gave difficulties:

> To see the odometer on a steep mountain trail is better fun than a circus; as it wobbles along a good road, it excites the curiosity and conjectures of the natives, to whom the one wheel without a body is the acme of ludicrous uselessness; but on a precipitous path, strewn with enormous bowlders and netted with chaparral, it shows the infinite possibilities of its motions. At one moment it bounds from the ground and saws the air; then it swings over the rider's head, and assumes the appearance of a patent hair-brushing machine of unusual proportions; and in extreme instances it reverses its normal motions entirely, and is propelled by the mule instead of dragging at that capricious creature's . . . tail.[23]

Lt. Hoxie had trouble with his odometer in southern Utah, remarking, "lost the use of the odometer today through fast driving, and I think the principle of construction faulty." A few weeks later, he notes "made 22¼ miles today by the odometer but as that wandered somewhat . . . I lost about a mile. I should call it 21 miles from Pelican Point to our camp in Draperville."[24]

The traverses were accomplished in generally north-south lines, with several teams following more or less parallel courses. At stations along the way, fixes were made on mountain peaks used in the triangulation, and the topographer in charge made contour sketches of the surrounding country. Travel was by horse back and by mule, and the care and feeding of the latter animals made for a constant, if not particularly welcome, diversion on the trail. Many a day was wasted in search of lost, stolen, or strayed animals. In November 1872, Lt. Hoxie recorded:

> Mules broke their picket lines last night and were missing this morning. Knew they were desperate for water and tied & hobbled all of them. Found them hobbled about 2 miles from camp.[25]

The final component of a topographic survey (and, for some scientific applications, the most important one) was determining the altitudes of points. The Wheeler survey took thousands of readings with barometers, meticulously corrected for temperature. Cistern barometers, while perhaps more accurate than the aneroid type, posed special problems in the field. During a dust storm in Colorado

> Lieutenant Morrison, of the Sixth Cavalry, our officer in charge, and Mr. Clark, the topographer . . .vainly attempted to fill a barometer tube with mercury. They worked patiently until after midnight, and then, after all their painstaking labor, a sudden gust of wind cracked and broke the slender glass.[26]

Later parties used aneroid barometers, but, on the whole, altitudes remained the most problematical part of Wheeler's field work. It is not unusual for his observations to be off by a hundred feet or more when compared with the elevations given on current Geological Survey maps.

Despite the exacting and time-consuming labor required for many of the measurements and observations, and the difficulty and sheer drudgery of much of the day-to-day movement, the Wheeler survey came to function like a highly efficient surveying machine, moving from south to north in its chosen regions. In its longest field season, during seven months in 1877, the survey rolled up impressive statistics, which Wheeler conscientiously, not to say tediously, re-counted in his final report. With 46 "professionals," five privates, an unnamed number of packers, herders, teamsters, and laborers, the survey was divided into six parties operating in Colorado, Nevada, and Utah. During that year they measured five base lines, occupied 106 main and 264 secondary triangulation stations, measured 10,800 miles on meanders, took over 10,000 barometer readings for altitude, made camp 761 times, collected 1,100 mineral and fossil specimens, 228 birds, 200 fishes, 11 lots of reptiles, 14 of insects, and 8 of shells.[27]

As these last items indicate, the field work of the survey did not end with mapping. Although the Wheeler survey stressed mapping more than any of the other great western surveys, most parties were accompanied by a geologist and one or two naturalists in various disciplines. The precedent set by the Pacific Railroad Surveys was difficult to shake off, and all four of the great surveys attempted more or less encyclopedic analyses of the regions in which they operated. There was great pressure from the scientific men of the age for a place in these federally funded surveys, perhaps the earliest cases of federal aid to science.

At various times the Wheeler survey employed the services of some forty civilian scientists in a dozen disciplines. The final reports have sections on geology, astronomy, paleontology, zoology, botany, archaeology, and ethnology. Although few of these scientists are remembered today except perhaps obliquely in the name of some western peak or butte, others made permanent contributions to their disciplines. Grove Karl Gilbert is perhaps the best example. As a young geologist with the Wheeler survey in 1871-73, Gilbert named and made the first scientific examination of the prehistoric Lake Bonneville, which covered most of Utah in pleistocene times. His first map of the reconstructed lake was published as a special atlas sheet by the Wheeler survey, and Gilbert went on to become a pre-eminent figure in American geology. Wheeler's photographer, Timothy O'Sullivan, joined the ranks of the immortals in his profession. Lieutenants Lockwood, Hoxie, Tillman, and others made valuable careers in the Corps of Engineers, and civilian topographers Louis Nell and Gilbert Thompson went on to influential work in private cartography and with the Geological Survey.

The strained relations between Wheeler and the civilian scientists under his (nominal) command have been often noted, and were a crucial factor in the eventual dissolution of his survey. The antipathy reflected a growing competition between military and civilian scientists. The time was not long past when scientific training meant a European university, or, in some fields, especially engineering, West Point. Formal training of civilian engineers in America dated from 1829, when the first class enrolled at Rennselaer Polytechnic Institute, and the scientific schools of Harvard and Yale did not open until the late 1840s.[28] The new American scientific community was experiencing growing pains, anxious to establish its *bona fides* in the international scientific community.

To James T. Gardiner, Clarence King's assistant, it was a point of honor that their survey, although under the War Department, was free from military interference. He stated

> I received my appointment on the geological survey of the 40th parallel from [Mr. King]. All of the other assistants were appointed by him after consultation with me. Whether these appointments had to be approved by General Humphreys [Chief of the Army Corps of Engineers] or not I do not know. We all felt that Mr. King was our leader, and that our allegiance was due him. It was well known to all of us that it was through his personal exertions and personal friends that the money for our support was appropriated by Congress. It made little difference to us what Department dispersed it so long as they did not meddle with our scientific plans.[29]

The military way was seen as hemmed in by tradition, bureaucracy, and subordination, and as antithetical to the new scientific ideals of untrammelled inquiry, individual expression, and intellectual innovation.

Geology had a special rank among scientific subjects and its ambitions were correspondingly large. Geology was on the one hand practical, "down to earth," with insights of immediate economic value to mining and agriculture. On the other hand, the vistas that it opened through its severely chronological approach to the earth's history made it an extremely powerful social force and a point of departure for other disciplines. Its subfield, paleontology, put geology right in the middle of Darwin's argument and gave theological implications to dry lists of strata. The last half of the nineteenth century has been called the heroic age of American geology, and it was Wheeler's misfortune to have to contend with its greatest heros, among them Hayden and Powell.

Wheeler was at pains to show that his relations with the civilian scientists under his employ were harmonious, and they appear to have been, at least outwardly, amicable and correct. He seems to have made a real effort to accommodate these somewhat prickly personalities and to allow them as much freedom as possible. In 1874 he wrote to the geologist John J. Stevenson "... I see no objection to the reduction proposed in your labors upon your geological report. There is no wish to limit you as to the time necessary for its completion."[30] Oscar Loew, a German mineralogist who had studied with the famous chemist Leibig, praised Wheeler for allowing him to pursue the then-new field of geochemistry: "... you have been the first to recognize the importance of chemistry as a branch of natural history operations in explorations for survey ..."[31]

Nor did the civilian scientists always live up to their responsibilities. Harvard Professor W. A. Rogers had been retained to do some of the mathematical corrections required in making latitude and longitude determinations. Rogers complained about his treatment by the survey and Wheeler answered "I regret that you should have undertaken the work unwillingly or feel that you have not been fairly compensated." He enclosed a memorandum by Dr. Kampf, a full-time civilian astronomer with the survey:

Before the work of Mr. Rogers, relating to the latitude of Cheyenne, was made up for the printer I controlled the computations as far as it could be done in the time and found a large number of mistakes in addition, the sign minus used often for the sign plus and vice versa .... The work seems to merit a careful overlooking before publishing.[32]

Gilbert's field work had been done in 1871-73, but in the latter year he met and came under the influence of Powell, whose training and personality made him a natural ally. Gilbert asked to be released from the survey so that he could join Powell. Wheeler, understandably, insisted that Gilbert first fulfill his obligations by submitting his written report. "I have ever sought," he wrote Gilbert, "to give to all the civilian specialists connected with the survey, the utmost scope for individuality in the prosecution of their duties and preparation of reports." Gilbert finally complied (though not before another dispute over the ownership of field books) and his reports were published by the survey in 1875.[33] E. D. Cope, the paleontologist, also chafed on the issue of publication, feeling that his discovery of fossils of early mammals in New Mexico required immediate airing in print, and fearing that each passing day would see him scooped by a French journal.[34]

A final aspect of field work not directly related to mapping was photography. Fremont had taken a daguerrotype camera west on his 1842 expedition, but never managed to get usable results,[35] and the Pacific Railroad Surveys seem to have relied solely on artists to provide their visual documentaton. All of the post-Civil War surveys, however, employed one or more photographers. The initial impulse was probably the scientific one of documenting the discoveries but the surveys soon discovered an even more powerful motivation for their photographs of western scenery: public relations. Wheeler went further in this direction than any of the other surveys. In what in retrospect seems a compulsive and doomed effort, he spent a good deal of his time and money publishing the work of his star photographer, Timothy O'Sullivan. O'Sullivan's photographs were made up into albums and stereopticon slides. A large volume of correspondence coming out of the headquarters office in Washington concerned the printing, mounting and distribution of these photographs. Many sets were distributed to institutions and public men, each set prominently identified with the name of the survey and Wheeler's name. A set went to Senator Logan of Illinois, who wrote asking for another for Illinois Normal University.[36] Professor James Dwight Dana wrote from Yale requesting twenty-five or thirty photographs showing geologic structure and valley erosion.[37] Not all of these publicity efforts fell on fertile ground. In debate in the House on survey appropriations in 1877, Representative Piper of California ventured to say:

> There is no gentleman on this floor who can point [to] one single item of advantage to the people or to the nation that has accrued from these explorations. It is true they take a great many stereoscopic views which are circulated about this House; in fact, I have got quite a box of them myself. They are very nice things for young gentlemen to amuse young ladies with, but I believe that is the only thing they are useful for . . . .[38]

In an early report on the Survey, Wheeler made a tantalizing allusion to photogrammetric methods, speculating that it would be possible "to give a value to the horizontal and vertical measurements upon a photographic picture," but I have seen no evidence that he ever followed up on the idea.[39]

The Survey's longest seasons were six or seven months, between May and December, but congressional delays in approving the War Department's budget sometimes retarded the schedule considerably. The 1874 season began on August 6th. Winter and Spring were seasons of equally feverish activity at the Washington office, as raw data were reduced to usable form, maps drafted, sent off for engraving and printing, distributed, reports written, and political fences mended.

Three to five draftsmen and two or three clerks were responsible for turning the year's data into maps. Draftsmen laid out the lines of latitude and longitude according to a modified secant conic projection with standard parallels at 34° and 44° north latitude. Data were entered on manuscript plotting sheets at the scale of 1 inch = 2 miles, using contours to indicate relief. The finished maps were drawn with pen and ink one-third larger than published size.

Although relief was plotted on the worksheets in the form of contour lines, these were converted on the finished drawing (and on the printed map) to hachures. The late nineteenth century was a period of experimentation in the depiction of relief and the four great surveys used three different techniques. King's maps used shaded relief and Hayden's and Powell's used contours. Wheeler's maps employed hachures, not of the "mathematical" type introduced by Dufour in Switzerland, where the width and spacing of hachures was keyed to the degree of slope, but "arbitrarily selected as to direction, number, and strength, allowance being made for light and shade, as best suits the scale."[40] Wheeler was fortunate in having as his topographical draftsman John Weyss, whose beautifully hachured maps drawn in the service of the Union army and of the Wheeler survey show absolute mastery of the technique. There is no denying that the survey's hachured sheets give a much clearer impression of the general terrain, viewed as a whole, than do contours. Unfortunately, the hachure is a technique more suited to the delicate, fine lines produced by copper engraving than to lithography, and in the printed maps of the survey, especially in mountainous areas of considerable extent, the hachures present a rather overwhelming mass of black, in which watercourses, cultural features, and lettering can only be picked out with difficulty.

Wheeler experimented with contours on a special map of the San Juan mining region in Colorado, issued in 1876 or '77. This sheet was innovative in that the contours below the tree line were printed in green ink and those above the tree line in reddish-brown ink. Hachures were used to highlight cliffs and escarpments, especially on the highest peaks. A plaster model of the area shown on this sheet was also shown at the Philadelphia Exposition in 1876.[41] Wheeler thought that by combining the two techniques, "the most striking effect is produced, the map is less obscured by heavy lines, and the more practical information, showing routes of communication, lines of drainage, settlements, etc., appears in bolder relief."[42] Given his propensity to catalogue the accomplishments of his survey in exhaustive detail, it is surprising that he does not make more of his introduction of this little refinement.

Another area in which Wheeler could justly claim a measure of innovation was in his series of land use maps. The first of these was issued in 1876, and in all, 27 land classification maps were produced. These constitute the first attempt at a uniform land classification map in the United States, an effort that has only recently been revived, after years of inactivity. Wheeler's maps were extremely simplified compared to modern land use maps which may differentiate dozens of categories of use. Wheeler's system identified four categories of land: 1) arable, 2) grazing, 3) timber, and 4) arid and barren. Even such broad information as this could have been useful for overall planning, for the guidance of homesteaders, and for the information of an army on the move. The land use maps were made by taking a photographic copy of an early state of an atlas sheet and instructing the topographers to mark on it "in colors, or . . . by other conventional signs, the land classification."[43] About one-quarter of the total area mapped by the survey was also published in the land classification series.

Field records were not the only sources used in compiling finished maps. The survey made use of published and manuscript maps from a variety of sources. Township plats of the General Land Office were

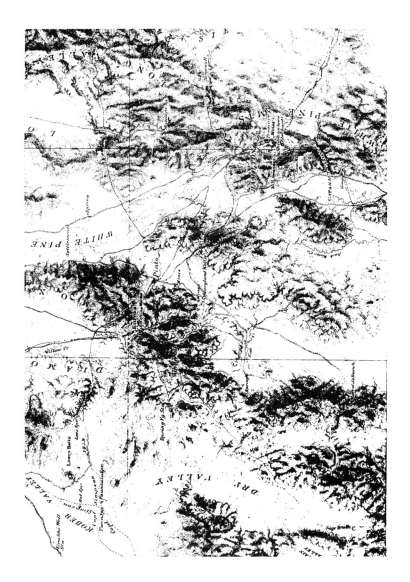

Fig. 21. Vicinity of Eureka and Hamilton, Nevada, from atlas sheet 49. Courtesy of The Newberry Library.

used as sources, as were published maps of other governmental and
private surveys. Wheeler borrowed a tracing of one of Powell's field
sheets which he used for the area north of the Colorado River on atlas
sheet 67.[44] He was careful to acknowledge Powell's help, as well as
data from Ives' 1858 expedition. This is the most derivative of
Wheeler's atlas sheets; he seems to have been in a hurry to include
the terrible canyon on his maps, probably as an illustration of the
locale of his own ill-conceived expedition. Other possible source
maps were requested or arrived unsolicited from railroad engineers
like General G. M. Dodge, and from publishers such as G. W.
Colton and O. W. Gray in New York, and B. A. M. Froiseth in Salt
Lake City. Wheeler invariably repaid the compliment by sending
some of his own printed atlas sheets in return, data from which the
commercial publishers were eager to include in their maps.[45]

In March 1873, Wheeler visited lithographic printers in Philadelphia and New York, seeking estimates for the engraving and
printing of his maps. He concluded that "Mr. Julius Bien of New
York was possessed of more facilities for executing the work in a
manner commensurate with its value, than any other establishment."[46] Bien was certainly not a surprising choice, for he had been
active in cartographic lithography for two decades and had done
much work for the government, which was to continue to use the
services of his firm extensively until Bien's death in 1909. Wheeler
monitored Bien's work carefully, and set the standards for paper,
color, layout and design, rejecting one paper sample because he
questioned its folding endurance. A flurry of correspondence passed
between Bien's New York plant and the survey offices in Washington. The greatest cause for concern was the accuracy of the engraving
of the lines of latitude and longitude. Wheeler wrote to Bien early in
1874 reminding him of "the artistic necessity of inspecting the lines
of junction of the several atlas sheets, for the true size of the
projection itself as well as for the fitting of all the details where the
sheets shall be joined together." This continued to give trouble. Two
months later he was writing "the proof of [sheet] 58 is so far out of
size that it will be impracticable to put in the corrections . . . .
Hereafter I think it will be well to have the stones sent here and the
projections laid off with the greatest accuracy or else send Mr. Nell

[the survey's draftsman] or someone to N.Y., whichever you think best."[47] As time passed, the difficulties were ironed out and a steady flow of manuscript maps, proofs, and printed sheets passed back and forth between Washington and New York, with "next day delivery" by express mail.

The office season was a time for preparing reports, doing research, planning the next field season; ordering printed forms, field notebooks, labels, and bookshelves for the office; ordering barometers, odometers, drawing paper and photographic supplies; corresponding with quartermasters at army posts about supplies, provisions, lost or mislaid vouchers, and (endlessly) about mules; and answering letters from job seekers and from public figures hoping to get their sons a summer job with the survey. There were frequent and urgent communications with Bien regarding the printing of the maps and with Chief of Engineers Humphreys about the financing of the survey. The bureaucracy of the time was not so bloated that Humphreys did not personally approve every expenditure including amounts of a few dollars. Interdepartmental requests were handled at the highest levels: Wheeler wrote to Ainsworth Rand Spofford, the Librarian of Congress, to ask if he could borrow Jukes' *Manual of Geology* and volume one of Whitney's *Geology of California* "for a short time for reference." A little more than a month later, Spofford wrote Wheeler, asking that they be returned.[48]

One of the most constant of the office activities, and one of the most likely to have a payoff in terms of public awareness and support of the survey, was the distribution of survey publications. This turned into almost a separate industry with atlas sheets and volumes of reports being sent out, gratis, by the thousands. The Survey was its own publisher, and Wheeler attacked the problem of distribution with characteristic energy and thoroughness. Meticulous records were kept of what was sent and to whom, and the lists are a who's who of influential men and institutions. The first fascicle of the topographic atlas, consisting of title page, legend sheet, progress map, a map of the western drainage basins, and atlas sheets 50, 58, 59 & 66 was distributed beginning on 21 April 1874. The first copy went to General Humphreys, followed closely by Speaker of the House James G. Blaine, chairman of the House Ways and Means and Appropriations committees Henry L. Dawes, Senator Roscoe Conkling of New York, Representative Benjamin Butler of Massachusetts, and Senator Morrill of Vermont. Somewhat further along in the list came the President, cabinet officers, the House Committee on Public Lands, numerous federal officials and Army officers, surveyors general of the states and territories, and domestic and foreign geographical societies. Wheeler had a list of colleges and went through it systematically, sending the maps to each, presaging the modern government document depository system.[49]

Later, up-to-date copies of the atlas were sent to virtually every mapping agency then functioning in the world, including the survey of India. These international contacts show that Wheeler was very conscious of the role of his survey in the general development of government topographical mapping, a theme that was to occupy him fully in his report on the International Geographical Congress in Venice in 1881. His was the only one of the four great surveys that was at all comparable to such venerable institutions as the Ordnance Survey of Great Britain or the Institut Geographique Militare, and Wheeler established personal contacts with the heads of several such agencies. To Field Marshal von Moltke in Berlin he wrote:

> Your attention is invited to the uniformity of size and scale, and the fact that the separate maps may be conjoined so as to comprise an entire political or other division,

adding somewhat apologetically,

> our government is yet young in its geographical undertakings, that have usually been the outgrowth of special wants and there is yet much to be perfected.[50]

The underlying antipathy between civilian and military cadres alluded to earlier came to the fore in a dramatic way during an encounter in Colorado in 1873. Hayden describes it:

> As we were riding down into the south Park, about the 9th day of July, we came across Lt. Marshall's party [of the Wheeler survey] and we camped together. He was a very courteous gentleman and we were very friendly. We talked matters over, and some regrets were expressed that we were on the same ground. I simply stated to him . . . that I had no option but to perform this

work, as we had had the Territory of Colorado assigned to us as a
field of exploration. He simply said that he was under orders, and
therefore could not disobey his orders.[51]

Two of the four surveys thus found themselves following the same
roads and trails and occupying the same mountain peaks, a situation
that, given the great discrepancies in their goals, should hardly have
been cause for undue alarm. Nevertheless, Hayden's ire was roused.
He had reportedly told an Army surgeon the previous winter "You
can tell Wheeler that if he stirs a finger, or attempts to interfere with
me or my survey in any way, I will utterly crush him — as I have
enough congressional influence to do so, and will bring it all to
bear."[52] His influence was at least sufficient to cause a congressional
committee to be set up to investigate overlap between the surveys. In
the spring of 1874, Wheeler, Hayden, and Powell were called before
Representative Townsend's Committee on Public Lands to give
testimony.

The committee was to investigate what surveys were operating in
the West, and whether they might not be consolidated under one
head or at least prevented from duplication of effort. During the
course of the hearings, Wheeler's survey was severely attacked on
grounds of inaccuracy and uselessness for geological work. Powell, in
his best professorial manner, seized on Wheeler's atlas sheet no. 59,
which covered an area in which Powell had worked:

> Professor Powell . . . explained inaccuracies with regard to
> Pangwitch Canon . . . also the failure of the map to show the
> proper topographic features at the head of the Sevier River and on
> the plateau to the east and that the map would contradict the
> necessary statement which would have to be made in discussing
> the geological structure of that region of the country. All this was
> explained by the use of diagrams on the blackboard.[53]

In similar chalk talks, Powell had described to the committee three
methods of executing topographical surveys, claiming that the most
accurate (triangulation from measured bases) was in use by Hayden
and himself and had been used by King, while the least accurate
method (meander survey) had been Wheeler's.

James Gardiner, now Hayden's chief topographer, faulted Wheeler's maps for their paucity of elevations. He maintained that there were shown on his published maps "great regions, as large as Connecticut and Rhode Island together, that he simply marched around and looked into, but did not enter," and pointed out discrepancies of "from one to two miles in longitude" between the match lines of two adjacent sheets.[54] J. D. Whitney wrote from Harvard "the recently published maps made under the direction of Lieutenant Wheeler seem to me very bad, and entirely behind the present requirements of geographical science in this country."[55]

Faced with such devastating reviews of his work, Wheeler might have made a number of telling points. He might have reminded the committee that the scathing criticism was all based on four published maps, deriving from the earliest work of the survey; that all of the surveys had begun with surveying techniques that they were later forced to upgrade, and that by 1874, there was very little difference in the quality of topographic work being performed; that his surveys were also beginning to use triangulation from measured base lines; that the high degree of accuracy claimed for Hayden's and Powell's maps could not be proven, since none of them had been published; and that the surveys were based on radically different premises. Actually, he did make some of these points in the course of a long and disputatious rebuttal, but he did not make them clearly or forcefully. Instead, he turned his remarks into an intemperate and personal attack, at times verging on the hysterical, on his least vulnerable opponent, Hayden.

He produced a notarized affidavit by H. C. Yarrow, an army surgeon on his survey, quoting Hayden's remark about crushing him if he got in Hayden's way. He questioned Hayden's knowledge of geodesy and topography and his ability to direct those parts of the survey; accused him of empire building "by jumping from one pretentious pinnacle to another," of benefitting unjustly from the researches of others; and of launching "a covert method of inquisitorial attack" upon his survey. Gardiner, he said, had practiced "deception," had been "traitorous" in leaving King's survey to work for Hayden, and had conspired to bilk the government out of a salary.[56]

Wheeler emerges in the Townsend hearings as hidebound and superior-ridden, attributes that could only have reinforced civilians' fears of military control of scientific surveys. Whether because of personal insecurity or official muzzling (one suspects the former), Wheeler's testimony has an almost paranoid quality; he frequently declined to answer questions, declaring that he was not in a position to know the answer or was following orders from above, or might be replaced at any time. His testimony has the disjointed, prolix, and vague characteristics that marked his prose. One can hardly doubt that he was technically very competent, but he lacked any ability to explain complex matters with at least the appearance of simplicity, an ability which Powell had, and used. The thirty-two-year-old lieutenant, lacking the patina of war service, and with an apparently rigid personality, must have contrasted badly with the self-assured, gray-haired Hayden and the cool, romantic, wounded veteran Powell.

Townsend noted that "ill-judged and hasty expressions have been used on either side, which good taste would have withheld,"[57] but finally recommended that all three surveys be kept in the field (King had concluded his field work and was busy with the final publications). In this way, it was thought, "a generous rivalry will be maintained among the good men therein, and a stimulus will be given to each to do the best work possible, and a resulting benefit will ensue in more accurate surveys and more extensive and valuable maps and reports."[58] The report also recommended the transfer of Powell's survey from the Smithsonian Institution to the Interior Department, a move that Powell fully concurred with.

But if the committee's findings gave a new lease on life to the three remaining surveys, it was to be a brief reprieve. The ruling was really a political decision to do nothing rather than a reasoned defense of the surveys, and the committee clearly felt that consolidation under the Interior Department was inevitable. The report had several immediate effects. Within a few weeks, the Interior Department had drawn up its own system for dividing the west into convenient-sized map sheets. They opted for a scale of four miles to an inch, covering the area west of 90° 30′ longitude in 146 sheets. There was to be a geographical and a topographical atlas.[59] This was clearly a response to Wheeler's systematic approach and Hayden and Powell made some effort to regulate their work according to the Interior Department sheet lines, but none of the maps were ever published and the grid was abandoned when the Geological Survey came into being. All parties probably took greater pains in their work. Powell's had a new staff member, G. K. Gilbert, one of whose first duties was assisting Almon Thompson, Powell's topographer. Gilbert immediately instituted some practices learned on the Wheeler survey, and passed on the manual of topographic procedure that Wheeler's Lt. Marshall had compiled.

The underlying issues, unresolved by the Townsend Committee, did not come to the fore again until 1878. Then another Congress raised almost the same questions as before, but with an added twist. Hoping to be able to pass the buck for what could not help but be an unpopular decision, the Congress invoked the aid of the National Academy of Sciences, an independent organization whose congressional charter gave it the responsibility of advising Congress, when requested, about scientific matters, particularly with regard to the costs of scientific research. The vexing problem of the surveys was presented to the NAS in June 1879 with the request that it be considered at the next meeting of the Academy. The NAS appointed a committee to report on the issue at its October 6th meeting. The committee did not spend much time in its deliberations; the chairman, Harvard paleontologist O.C. Marsh, asked the heads of all the surveys (including the General Land Office surveys of public lands) to send statements to the committee. The replies of the survey chiefs were all dated between October 28th and November 1st, and the committee met on November 5th. The next day the committee's report was approved by the membership of the Academy.[60]

Given the makeup of the committee, all academic scientists, there could have been little doubt that the army would fare poorly in its report. Actually, the report rejected all the existing survey organizations, but did advocate Interior Department control. The course proposed was a radical one that made good sense from a logical point of view. There would be three organizations working together within the Interior Department. The first, a reorganized Coast and Geodetic Survey (renamed the Coast and Interior Survey) would be responsible for all "surveys of mensuration" — geodetic, topographic, and land parcelling surveys. Secondly, there would be a new organization, the U.S. Geological Survey, to study "geological structure and economical resources" including land classification and valuation. Finally, the General Land Office, shorn of all its surveying responsibilities, would be charged with the "disposition and sale of public lands, including all questions of title and record." It was understood that both the Geological Survey and the General Land Office would have to issue special maps to illustrate their work, but these would always be done using base maps prepared by the Coast and Interior Survey. There was no place for the military in this scheme, except that "officers of the Army and Navy, when not otherwise employed, might be detailed by the Secretary of War, or of the Navy, to take part in the operations of the general survey." A final, revolutionary clause called for a committee, consisting of the heads of the three new and reorganized departments, the Chief of Engineers of the army, and three others appointed by the president, to devise "a standard of classification and valuation of the public land, together with a system of land-parcelling survey."[61]

The last provision was probably instigated by Powell, who would elaborate on it in his book on the arid lands published in 1879, but no one else was ready to try to dismantle the deeply ingrained township and range system, and no one else was willing to try to survey the public lands with the level of geodetic control the proposal envisioned.

The army's reply to the NAS report stressed its fiscal irresponsibility. Contrary to the instructions of their charter, the NAS had not considered economic factors, and Humphreys demonstrated that the proposed survey, carried out with sufficient refinement to enable it to be used for cadastral purposes, would have been extremely expensive, on the order of $403 per sq. mile. This was contrasted with Wheeler's liberal estimate of $2.50 per sq. mile to finish his maps, which he thought could be accomplished by 1893.[62]

Congress was only partly convinced. In the spring of 1879 it created the U.S. Geological Survey which would assume all surveying, mapping, and geological investigations in the West. The General Land Office, however, would continue to survey public lands in the time-honored way, and the Coast and Geodetic Survey (still under the Treasury Department) would continue to do first-order triangulation for the whole country.

Appropriations for field work for Hayden's, Powell's and Wheeler's surveys ended on 30 June 1879, the month that the NAS committee had been appointed; the appropriations were never renewed. Wheeler continued to publish the final reports of his surveys as well as maps for which the field work had been completed, and the office of the Geographical Surveys West of the 100th Meridian did not close until 1884. But Wheeler was broken in health and spirit. He was on leave from the army for reasons of health for many months in the 1880s and retired with a disability in 1888. After the final volume of his report (actually, volume one) was published in 1889, he resigned his commission and virtually dropped out of sight. He went into private engineering practice and apparently maintained residences in Washington and New York.[63] In 1893 he published a scheme for the organization of an international exposition, and sent a copy to William Frederick Poole, the first librarian of the Newberry Library, wondering whether the scheme might be useful in organizing a library's collections.[64]

Wheeler died in New York City in 1905, his passing recorded in a one paragraph obituary in *The New York Times* which was reprinted, unchanged, in the *Army and Navy Journal*.[65] He has remained the most obscure of the figures associated with the four great surveys, and his work has dropped most completely into oblivion.

But Wheeler is not entirely without his monuments. The most visible and permanent is Wheeler Peak in eastern Nevada, which the young lieutenant had the good sense to name early in his career. Others were of the human kind, the dozens of scientists and engineers who went on to build substantial careers on lessons learned with the Wheeler survey. But Wheeler's most important contributions were the topographic atlas sheets he published, the manifestations of what he called "a connected survey." It is a measure of the usefulness of this work, and the lack of subsequent interest in compiling a general topographic map of the country, that some sheets of the Wheeler atlas were still being sold by the Superintendent of Documents as late as 1913.[66] The U.S. Geological Survey

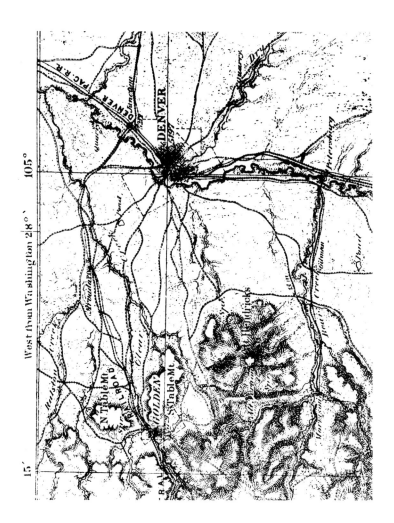

Fig. 22. Vicinity of Denver from atlas sheet 53C.
Courtesy of The Newberry Library.

eventually published maps of the entire area west of the 100th meridian at the scale Wheeler had used — the first such sheet was published in 1945 and the last in 1965. And ironically, the maps were actually made, not by the Geological Survey, but by the Army Map Service.

Historians and historical geographers in particular can lament the premature passing of the Wheeler survey, because with it went our last, best chance to have the cultural features of the old west at the close of the frontier era systematically recorded for posterity. At the close of his survey, Wheeler had mapped about one-quarter of his chosen area, and could almost surely have completed all the mapping by 1893, as he claimed. With roads, trails, telegraph lines, settlements, and other such ephemera marked from Puget Sound to the Rio Grande and from Catalina Island to Dodge City, what a resource the western historian would have had!

NOTES

1. William H. Goetzmann, *Army Exploration in the American West* (1959), p. 313.

2. George M. Wheeler, *Report upon the United States Geographical Surveys West of the 100th Meridian* (1875-1889), vol. 1 (1889), p. 763. The final report of the Wheeler survey, in seven volumes.

3. Ferdinand V. Hayden, *Geological and Geographical Atlas of Colorado and Portions of Adjacent Territory* (1877).

4. Clarence King, *Geological and Topographical Atlas Accompanying the Report of the Geological Exploration of the Fortieth Parallel* (1876).

5. Richard A. Bartlett, *Great Surveys of the American West* (1962), p. 212. The best general account of the four surveys.

6. Wheeler, *Report,* vol. 1 p. 717.

7. *Map of Reconnaissance through Southern and Southeastern Nevada* (1869).

8. The first attempt at a multi-sheet topographic map of the United States was made by Christopher Colles in 1794. Only

five sheets were published; see Christopher Colles, *A Survey of the Roads of the United States of America,* ed. by Walter W. Ristow (1961), p. 77-81.

9. Bartlett, *Great Surveys,* p. 334.

10. Robert M. Utley, *Frontier Regulars: The United States Army and the Indian, 1866-1891* (1973), p. 23.

11. Richard L. Hoxie Diary, 16 October 1872, Hoxie Papers, Manuscripts Division, Library of Congress.

12. Samuel Escue Tillman Memoir, p. 54, Archives and Manuscripts Section, Tennessee State Library and Archives.

13. Wheeler, *Report,* vol. 1, p. 170.

14. Stephen J. Pyne, *Grove Karl Gilbert: A Great Engine of Research* (1980), p. 61.

15. Wheeler, *Report,* vol. 2 (1877), p. 86-96.

16. Henry Gannett, *A Manual of Topographic Methods* (1893), p. 19-20.

17. Wheeler, *Report,* vol. 1 p. 334.

18. William H. Rideing, "The Wheeler Survey in Nevada," *Harper's New Monthly Magazine* 55 (1877): 68.

19. William H. Rideing, *A-Saddle in the Wild West* (1879), p. 77-78.

20. House Committee on Public Lands, *Geographical and Geological Surveys West of the Mississippi,* H. Rep. 612, p. 51, 43d Cong., 1st Sess., 1874, *Congressional Serial Set* no. 1626. Hereafter cited as Townsend Report.

21. Gannett, *Manual,* p. 22.

22. Wheeler, *Report,* vol. 1, p. 347-8.

23. Rideing, "The Wheeler Survey in Nevada," p. 68.

24. Hoxie diary, 14 November and 5 December 1872.

25. Hoxie diary, 19 November 1872.

26. Rideing, *A-Saddle,* p. 20.

27. Wheeler, *Report,* vol. 1, p. 685.

28. Goetzmann, *Army Exploration,* p. 13.

29. Townsend Report, p. 70.

30. Wheeler to Stevenson, 7 February 1874, Press Copies of Letters Sent, Item 363, Records of the Office of the Chief of Engineers, Record Group 77, National Archives (hereafter RG 77, NA).

31. Wheeler, *Report*, vol. 3, p. 571.

32. Wheeler to Rogers, 13 February 1874, Item 363, RG 77, NA.

33. Wheeler to Gilbert, 7 June 1874, Item 363; Gilbert to Wheeler, 30 October 1874, Digests of Letters Received, Item 366, RG 77, NA.

34. Cope to Wheeler, 13 December 1874, Item 366, RG 77, NA.

35. Charles Preuss, *Exploring with Fremont: The Private Diaries of Charles Preuss, Cartographer . . .* , trans. and ed. by Erwin G. and Elisabeth K. Gudde (1958), p. 32-33, 35.

36. Logan to Wheeler, 14 March 1874, Item 366, RG 77, NA.

37. Dana to Wheeler, 19 February 1874, Item 366, RG 77, NA.

38. 21 February 1877, *Congressional Record*, 44th Cong., 2d Sess., House, p. 1793.

39. George M. Wheeler, *Progress-Report upon Geographical and Geological Explorations and Surveys West of the One Hundredth Meridian, in 1872* (1874), p. 11.

40. Wheeler, *Report,* vol. 1, p. 398.

41. Ibid., p. 270, note.

42. Ibid., p. 398.

43. Wheeler to Louis Nell, 27 October 1878, Item 363, RG 77, NA.

44. Wheeler to J.W. Powell, 21 April 1874, Item 363, RG 77, NA.

45. Dodge to Wheeler, 11 March 1876; Froiseth to Wheeler, 6 February 1874; Gray to Wheeler, 29 April 1874, Item 366; Wheeler to Colton & Co., 4 May 1874, Item 363, RG 77, NA.

46. Wheeler to Gen. A. A. Humphreys, 31 July 1874, Item 363, RG 77, NA.

47. Wheeler to Bien, 20 March, 2 May 1874, Item 363, RG 77, NA.

48. Wheeler to Spofford, 5 March 1874, Item 363; Spofford to Wheeler, 24 April 1874, Item 366, RG 77, NA.

49. Distribution Record of Reports, Memoranda, and Atlas Sheets, Item 388A, RG 77, NA.

50. Wheeler to von Moltke, 4 June 1874, Item 363, RG 77, NA.

51. Townsend Report (see note 20), p. 32.

52. Ibid., p. 63.

53. Ibid., p. 51.

54. Ibid., p. 57.

55. Ibid., p. 62.

56. Ibid., p. 67.

57. Ibid., p. 18.

58. Ibid., p. 17.

59. Secretary of Interior Delano to J.W. Powell, 1 July 1874; printed in H. Ex. Doc. 80, 45th Cong., 2d Sess., 1877-78, *Congressional Serial Set* no. 1809.

60. *Letter from O.C. Marsh . . . Transmitting . . . the Report on the Scientific Survey of the Territories, Made by the National Academy of Sciences,* S. Misc. Doc. 9, 45th Cong., 3d Sess., 1878-79, *Congressional Serial Set,* no. 1833.

61. Ibid., p. 2-4.

62. *Letter from the Secretary of War, Communicating Information in Relation to the Surveys in the Territory West of the Mississippi River . . .* , p. 15; S. Ex. Doc. 21, 45th Cong., 3d Sess., 1878-79, *Congressional Serial Set* no. 1828.

63. Peter L. Guth, "George Montague Wheeler: Last Army Explorer of the American West." Thesis, United States Military Academy, 1975, p. 63-64.

64. Wheeler to Poole, A.L.S., 20 October 1893, bound into the Newberry Library copy of Wheeler, *A Universal World's Exhibit* (1890).

65. *New York Times,* 5 May 1905, p. 9; *Army and Navy Journal,* 6 May 1905, p. 969.

66. Accession note on a map in the American Geographical Society Collections, Golda Meir Library, University of Wisconsin — Milwaukee.

MAPPING THE UNITED STATES-MEXICO BORDERLANDS: AN OVERVIEW

Norman J.W. Thrower

University of California
Los Angeles

A good place to begin this

survey of the mapping of the southwest of the United States might
be to review, briefly, the broad outlines of the modern geography of
this generally arid and semi-arid area. (fig. 23) El Paso, Texas,
roughly in the middle of the 3,000 kilometer long boundary is the
pivot or fulcrum. Eastward of this city the international boundary is
formed by the Rio Grande, or the Rio Bravo del Norte as it is known
in Mexico. This great river rises in the 4,000 meter high Rocky
Mountains of the United States and flows generally southward past
the cities of Albuquerque and Las Cruces, New Mexico, to El Paso.
Santa Fe is on a stream of the same name which is a tributary of the
Rio Grande; it is about eighty kilometers upstream from the main
river, and northeast of Albuquerque. At El Paso the direction of the
Rio Grande changes to generally south-eastward, except at the Big
Bend where, for a hundred kilometers, the direction is northeast. The
Rio Grande reaches the Gulf of Mexico at a classic delta on which
the twin cities of Brownsville, Texas, and Matamoras, Mexico, are
situated.

To the west of El Paso the international boundary is largely
artificial, as indicated by the straight lines of modern instrumental
surveys, except for a short stretch of the Colorado River. This
western, dryer section, which is the main focus of this paper,
alternates between mountains up to 3,000 meters high and broad
basins; the lowest of these basins, the Salton Trough, is below sea
level. This arid area is drained by the Colorado River, which empties
into the Gulf of California, and by its tributaries. For our consideration, the most significant of these tributaries is the Gila River, whose
confluence with the Colorado is just north of the international
boundary. The most important cities in this drainage area are Tucson
and Phoenix, Arizona, and Mexicali, Mexico. At the coast is the Los
Angeles and San Diego urbanized area of some ten million inhabitants, now the largest metropolis on the Pacific coast of the Americas,
from Alaska to Tierra del Fuego, Chile. Los Angeles, the second

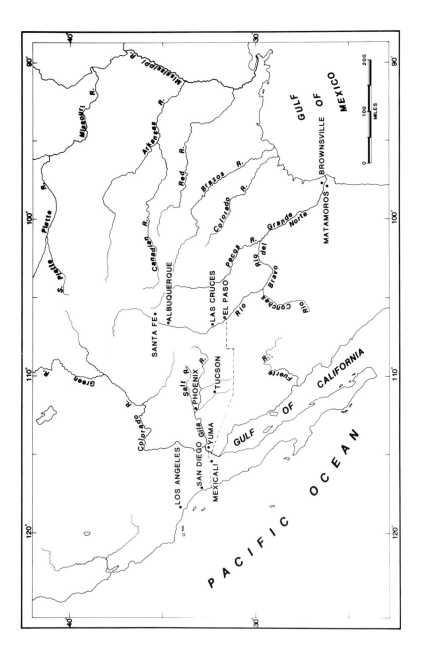

Fig. 23. Location map of the United States — Mexico Borderlands.

Mexican City in terms of origin of inhabitants after Mexico City, has a Mediterranean type of climate, with predominantly low sun (winter) precipitation. Elsewhere, the climate grades from a true desert in the interior, to steppe climate in west Texas, to sub-humid at the Gulf Coast. The present international boundary is a border across which there is a constant flow of ideas, goods, and people.

EARLY GENERAL MAPS OF THE BORDER

Even before the conquest of Mexico by Hernándo Cortés in 1521, exploration began along the coast of the Gulf of Mexico (fig.24). In 1519 Alonzo Álvarez de Piñeda was in the area of the Rio Grande delta followed, less than a decade later, by the coastal charting exploration of Panfilo de Narváez. A member of this expedition, Álvar Núñez Cabeza de Vaca, survived shipwreck on the Texas Coast and, with three companions, went overland by way of the Big Bend of the Rio Grande to Mexico's west coast. The party reached Mexico City in 1536 and, the next year, Cabeza de Vaca returned to Spain where he spread stories of the rich frontier land.

Meanwhile Cortés, who had been replaced in Mexico City by a Viceroy, Antonio de Mendoza, began exploring the coasts of the Gulf of California. There was thus a two pronged attack on the frontier: largely by sea on the coast of the Californias, supported by Cortés; and northward by inland routes sponsored by the Viceroy. Cortés' lieutenant, Francisco de Ulloa reached the head of the Gulf of California in 1539 and then sailed around the peninsula to 28° north, on the Pacific side. At about the same time, Hernándo de Alarcón ascended the Colorado River from its delta almost to the present international boundary. Another of Cortés' lieutenants, the Portugese-born, Juan Rodríguez Cabrillo sailed along the Pacific coast past the present international boundary in 1542.

Contemporaneous with these coastal reconnaissances were overland explorations sponsored by the Viceroy from Mexico City. In 1539 a Franciscan padre, Marcos de Niza travelled north crossing the Gila River to a Zuni pueblo; he said he saw this only from a distance before retreating, because of the hostility of the Indians. From them he learned that it was called Cíbola or Cívola; this and other pueblos

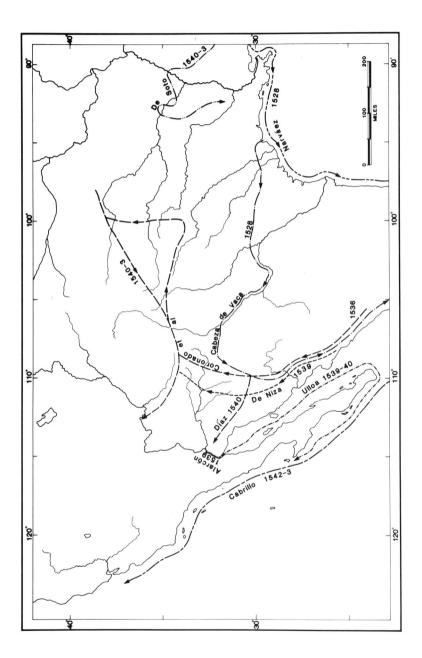

Fig. 24. Routes of Major Spanish Explorers in New Spain, 1528-1543.

which he named were soon recorded on the maps of Sebastian Munster, Batista Agnese and Giacomo Gastaldi. Gastaldi's map of 1546, especially, exemplifies the cartography of the borderlands based on reports of travellers.

De Niza's expedition was shortly followed by another, also sent out by the Viceroy Mendoza, under the leadership of a soldier, Francisco Vásqez de Coronado. Coronado attacked and took Cibola but, to his great disappointment, found no gold. Coronado set up his headquarters near present-day Albuquerque and sent expeditions in all directions. The most important of these for our story is the expedition of Melchor Díaz who went down the Colorado River to a point visited a year earlier by Alarcón. Díaz thus passed the confluence of the Gila and Colorado Rivers. After Coronado's return to Mexico City in 1542 no more expeditions were sent northward for about forty years. But the maps of Agnese and Gastaldi, from the mid 1550's specifically refer to the explorations of Coronado.

The contemporaneous explorations of Hernando de Soto in the eastern United States, and those of Coronado in the western part of the country enabled the cartographers to express on their maps something close to the correct width of what is now the United States. This is true of the influential maps of Gerhardus Mercator and Abraham Ortelius of the second half of the sixteenth century. In spite of this, maps continued to be published throughout the seventeenth century which showed a continent of very limited longitudinal extent.

RECONNAISSANCE MAPS — 17TH AND 18TH CENTURIES

What amounted to a re-discovery of New Mexico from the south was undertaken by padres and soldiers at the end of the sixteenth century. The earliest extant European map made in the field resulted from the expeditions of Juan de Oñate made after 1594; it was published in 1602. Oñate was appointed Governor General of New Mexico with the understanding that he would finance his own expeditions. In 1598 Oñate crossed the Rio Grande at El Paso and proceeded up the river, visiting pueblos on the way. Oñate's route is shown on his sketch map, on which there are notations concerning

Indians, towns, etc. and a statement to the effect that "there is gold but none of our men saw it or found any trace of it." Like Coronado, Oñate sent out expeditions from his headquarters, San Juan (fifty kilometers north of Santa Fe) — one of these was to the extremely dry Southwest which was judged to be a most unpromising area.

A new Governor, Pedro de Peralta was appointed in place of Oñate in 1608 and it was he who established Santa Fe as the capital of the province. Santa Fe was reached from the south by way of the upper Rio Grande, Oñate's route, by generations of Spanish padres, soldiers, traders and others throughout the 17th and 18th centuries.

During this period one of the most remarkable examples of what R.A. Skelton called "retrogression" in cartography took place in respect to California. In general, maps of the sixteenth century, such as those of Ortelius, showed California to be firmly attached to the mainland — then, at the beginning of the 17th century, it began to be represented as an island. This geographical myth was only finally overthrown in the early 18th century. The man chiefly responsible for this was Father Eusebio Kuhn (in Spanish, Kino), who spent fifteen years in the area of the present California/Baja California boundary as a missionary before he had enough evidence to delineate California as once more attached to the mainland. He did this on his map, "Passo por Tierra a la California," 1701.

During the 18th century maps were made in the field, of various localities of the frontier, by engineers, administrators and padres, which improved the geographical knowledge of these smaller areas. For example, between 1725 and 1729 a trained engineer Francisco Alvarez y Barreiro, drew the best map of the upper Rio Grande settlements that had been made up to that time. He also gave cartographic recognition to the myth that the Aztec nation had its origin at a lake at the headwaters of the Azul, a tributary of the Gila River. In the middle of the eighteenth century Francisco Antonio Marin del Valle sent a pictorial map to the King of Spain which included the names of Indian tribes, together with substantial statistical data being called for at that time from the provinces. A Spanish engineer Nicholas Lafora co-authored a map of the Frontier of New Spain in 1771, on which the Gila and Upper Rio Grande drainage is better shown than on any previous map.

Among the maps of California of the later 18th century those of Father Pedro Font are perhaps the best known. One of Font's maps, dated 1776, delineated a journey he took from Sonora, to just north of San Francisco. At the same time the Spanish entered San Francisco Bay from the sea. This was at the beginning of the mission period in Alta California which had been anticipated by the founding of missions in New Mexico more than a century earlier.

GENERAL MAPS OF THE EARLY TO MID 19TH CENTURY

The most noteworthy general map, up to its time, embracing what was to become the United States borderlands and a wide area beyond is that of Alexander von Humboldt. Humboldt first developed his map of New Spain in 1804, completed it in 1809 and had it published in 1811. He consulted many sources including the maps of Kino, Berriero, Lafora, Font, and others. Because of Humboldt's international stature it became the standard map of the area for nearly half a century.

During this period Mexico freed itself from Spain, in 1823, but lost Texas to the Americans in 1836 and New Mexico, Arizona, and California in the Mexican War, 1846-1848. The Americans had been penetrating the area since the very early years of the century when Zebulon Pike, in 1806, entered New Mexico and Texas from the north and west. (fig. 25) Pike was arrested and then released by the Spanish authorities; his publication sparked much interest in the southwest, in the United States. The travels of the fur trader and mountain man Jedediah Strong Smith, in California, after he had visited that area in 1826-7, likewise excited the interest of Americans.

A number of general maps were made by United States Army officers and by American commercial map publishers at mid-century. An example of the first is provided by the work of William Hemsley Emory. Emory had been trained in surveying at West Point and was commissioned First Lieutenant in the Corps of Topographical Engineers in 1836. His first experience in compiling maps from original data came when he prepared Joseph Nicollet's map of the Upper Mississippi for publication. For his role in this important work it has

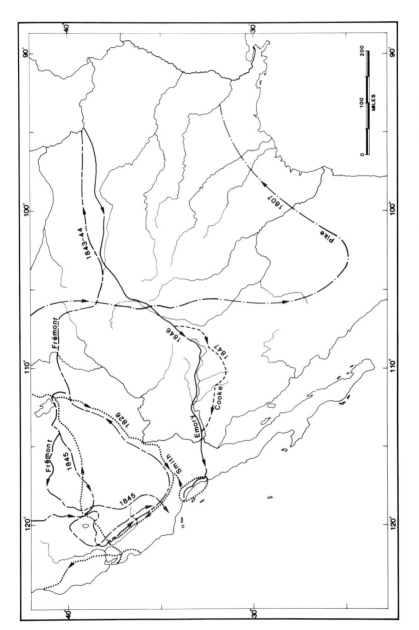

Fig. 25. Routes of Major American Explorers in the West, 1806-1848.

been said that it was Emory, rather than Nicollet's protege John Charles Fremont, who became Nicollet's last true disciple. After this experience Emory made a map of Texas, in 1844, showing the claims of that state. For this he used all available sources including Humboldt's map.

An example of American commercial cartographic publishing of about this time is the map by John Disternell, dated 1847. The next year this was used as setting forth the claims of the United States, vis-a-vis Mexico. It was used as the basis of the Treaty of Guadalupe Hidalgo which ended the Mexican War in 1848.

MAPS FROM INSTRUMENTAL SURVEYS

Emory, an outstanding mathematician and cartographer, was now ready to make his important contributions to the original mapping of the borderland. In 1846, at the outbreak of the United States War with Mexico, Emory was made the head of a topographic unit attached to General Kearney's Army of the West. Using barometers, chronometers and sextants Emory and his party surveyed the Upper Rio Grande and Gila Rivers and proceeded to California. At San Diego, which he reached in December 1846, Emory was able to verify his longitude with that taken earlier by Captain Sir Edward Belcher of the British Royal Navy.

A little later, Colonel P. St. George Cooke made a reconnaissance south of the Gila River through Tucson. Cooke's route was plotted on Emory's map, published in 1847. With Emory's recommendation this was, later, the basis of the Gadsden Purchase, December, 1853. Emory's map and his report are contemporaneous with those of his rival Fremont, generally to the north. Both are the forerunners of those great post-Civil War surveys of the American west which take their names from their leaders — King, Wheeler, Hayden and Powell (of the late 1860's and the 1870's).

A major detailed mapping project following the defeat of Mexico in 1848 was the United States Mexican Border survey. For political reasons John Russell Bartlett, who was totally unqualified for the work, rather than Emory, was made the head of the United States Boundary Commission. Eventually Emory was assigned as chief astronomer in charge of running the boundary line between the United States and Mexico. This work took from 1848 to 1853.

As head of a newly-created Bureau of Exploration and Surveys, Emory also laid out initial routes for southern transcontinental railroads to be built later. The Pacific Railroad Reports 1855-1860, with an accurate map on the scale of 1:3,000,000 by Lieutenant G. Kemble Warren, resulted from these surveys, but it was Emory who began the project and developed the methodology.

LATER DEVELOPMENTS

Following the Civil War and the establishment of the United States Geological Survey in 1869, detailed topographic maps were made of various areas on the border, as in other parts of the United States. As elsewhere, at first these maps were made by conventional ground survey methods but, after the development of aerial photography following World War I, by photogrammetry.

After World War II and the initiation of space flight, the southwest of the United States became a prime area for experimental imaging. Gemini and Apollo flights produced excellent, but ad hoc, coverage of the area. The launching of Landsat I, in 1972, permitted continuous, repetitive coverage by imagery of this and most of the rest of the world. Later Landsat missions, in concert with earlier ones, enable us to view any particular area every nine days. Thus, after over four centuries of cartographic endeavor in the area, we are now close to the ideal in possessing a uniform, repetitive and synoptic view. The ideal will have been reached when these qualities are combined with images of greatly improved resolution of the United States/Mexican Borderlands and all areas of the earth.

Maps referred to in the text can be found in original, facsimile or other reproduction in the following works.

Martha C. Bray, *Joseph Nicollet and his Map* (Philadelphia: American Philosophical Society, 1980).

William H. Emory, *Memoir to Accompany the Map of Texas* (Washington: United States Senate Document Vol. V, No. 341, 1844).

——————————, *Notes of a Military Reconnoissance* [sic] ... *Made in 1846-7* (Washington: United States Senate Executive Document No. 7, 1848).

——————————, *Report on the United States Mexican Boundary Survey* (Washington: United States Congress Executive Document No. 135, 1857).

George W. James, *Francisco Palou's Life ... of ... Junipero Serra (Pasadena, 1913).*

John B. Leighley, *California as an Island: An Illustrated Essay with Twenty-Five Plates and a Bibliographical Checklist of Maps ... 1622-1785* (San Francisco: Book Club of California, 1972).

James C. Martin and Robert S. Martin, *Contours of Discovery: Printed Maps delineating the Texas and Southwestern Chapters in the Cartographic History of North America* (Austin: Texas Historical Association and The University of Texas, 1981).

Dale L. Morgan and Carl I. Wheat, *Jedediah Smith and his Maps of the American West* (San Francisco: California Historical Society, 1954).

Seymour I. Schwartz and Ralph E. Ehrenberg, *Mapping of America* (New York, 1979).

Ronald V. Tooley, ed. *California as an Island* (London: The Map Collector's Circle, 1964).

Henry R. Wagner, *Cartography of the Northwest Coast of Amercia to the Year 1800* (Berkeley: University of California, 1937).

Carl I. Wheat, *Mapping the American West* (Worcester, Massachusetts: American Antiquarian Society, 1954).

——————————, *Mapping the Transmississippi West, 1540-1861, 5* Vols. (San Francisco: Institute of Historical Cartography, 1957-1963).

MAPPING THE SOUTHWEST: A TWENTIETH CENTURY HISTORICAL GEOGRAPHIC PERSPECTIVE

John B. Garver, Jr.

National Geographic Society
Washington D.C.

The National Geographic

Society's new historical map series, "The Making of America," is an ambitious five year project consisting of 17 regional maps depicting the historical geographic development of the United States from pre-Columbian days to the present. Announcement of the series appeared in the November 1982 issue of the *National Geographic* which introduced the first map of the series, *The Southwest*.

On the following pages three topics are addressed. First, a brief definition of historical geography, second, a discussion of the historical geography approach to mapping the Southwest, and third, a review of the process of making this map.

While writings combining history and geography trace back many centuries, the decades of the 1940s and 1950s saw the emergence of historical geography as a recognized subfield of the discipline of geography in the United States mainly through the writings of Ralph H. Brown of the University of Minnesota and Andrew H. Clark of the University of Wisconsin. Today, the study of historical geography has many able scholars, a number of whom are contributing to the preparation of our map series on "The Making of America." Historical geography may be viewed as the reconstruction of past geographies and the interpretation of geographical change through time. Thus, the emphasis of historical geography is upon *places* rather than *persons*. People are obviously an important part of most places, but the focus is not upon them directly, but how they have affected or given character to places. Through time a place or "region" becomes stamped or marked with a distinctive man-made or cultural landscape.

Specialized knowledge, tools, and techniques are needed to study and record past cultural landscapes — knowledge of both physical and human geography, of the history of the area, of cartography, air photo interpretation, and field observation. The mastery of document and archival research are essential elements as are skills in geographic description and historical narration. Special products of such studies are maps which graphically display the diverse areal patterns that have formed within the region through time.

Map portrayal of the historical geography of the Southwest United States includes a great variety of elements and themes which combine to reflect the changing man-land experience as different culture groups over time have interfaced with the land and with each other. These elements of change have produced successive human geographies in the Southwest — each made up of a composite of the old and the new where the pressures of an intruding dominant culture upon a weaker regional culture have transformed both, observable through time in the evolution of a modified cultural landscape.

In the map presentation of the historical geography of the Southwest more than a thousand years of occupance by various peoples had to be shown in an organized, accurate manner and on a sheet of paper measuring only 20½ X 27½ inches — a medium that must reflect effectively the old saw that "A picture is worth a thousand words." This is where the continuing challenge of the project takes place as a map of each of the 17 regions is designed and produced. Consider if you will the themes of the series: peoples, economics, settlements, networks, and resulting cultural landscapes. These themes had to be examined for each of the several cultural groups of the Southwest using the principal forms of map display — points, lines, and areas. (Map 26).

In presenting the diversity of the historical geography of the Southwest, there are three elements in our approach: illustrations (art), text, and of course, maps. Each contributes to the telling of the story. Through illustrations and paintings, the diversity of the peoples can be depicted as can the evolution of technology and the reconstruction of past physical and cultural landscapes. On the Southwest map we can see settlement types ranging from ancient cliff houses to modern communities, different peoples ranging from Pueblo Indians to conquistadors, priests, and cowboys, and technology from a prairie schooner to a railroad locomotive. The narrative on the map contributes to telling the historical geography of the region by presenting information not readily conveyed in graphic form.

Let us now consider the Southwest as presented on the first map of the "Making of America" series. Just what were these dynamic, changing geographies of the past and how did they evolve and even endure as a part of today's Southwest landscape? There were three

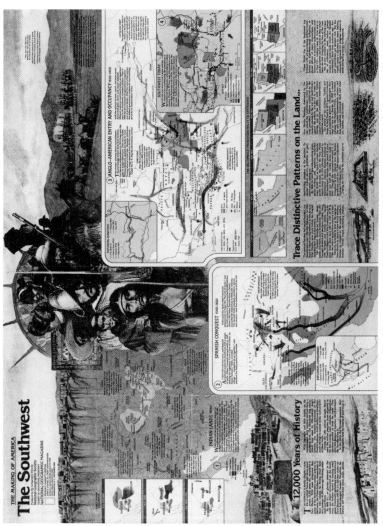

Fig. 26.

peoples or cultures that arrived at different times in the region: the
Indians, the Spanish and the Anglos. Through time a sequence of
cultural landscapes evolved — each succeeding one a blend of the old
and the new. The nature of the occupancy and the change is better
documented for some culture groups and eras than for others. It
ranges from observable ruins of early Indian presence to maps,
reports, sketches, and photographs of the more recent Spanish and
Anglo intrusions.

A useful observation which reflects the physical stage of the
historical geography of the Southwest is the harsh environment of the
region. This is depicted on the maps by the tan, yellow, and light
green color tints on both map sides and by the rugged relief pattern
for mountainous areas on the contemporary map side. The evolving
cultural landscapes are mainly confined to settlement nodes near
water and to the links between nodes that trace rivers and streams.
The physical feature most central to the historical geography of the
Southwest is the river with its water, its fertile valley, and its crossing
points. The river valleys provided sites for settlement and routes for
internal and trans-area movement across the land.

The first map grouping entitled "Early Indian Occupancy"
A.D. 600 — A.D. 1450 depicts semi-sedentary peoples who practiced hunting, gathering, and rudimentary agriculture. These poeples
included the Anasazi, Hakataya, Hohokam, and the Mogollon.
Through the map sequence we can see a variation of locations as
some movement and mixing occurred. The rugged environment,
frequent drought, the spread of agriculture, and the impact of the
nomadic Paiutes and Athapaskans are key elements of change in this
early geography. Knowledge of the period comes to us through the
field work and artifact studies of archaeologists and ethnohistorians.

The "Indians Lands" ca. 1600s map portrays a changing cultural
landscape of rancheria, pueblo, and nomadic peoples. The rancheria
peoples are in scattered valley farming settlements, nomadic tribes
have spread throughout the mountain fringes from which they raid
their sedentary neighbors. The Pueblo Indians occupy permanent
adobe villages along the Rio Grande and tributaries of the Little
Colorado River. Primary sources for this map are limited to journals,
reports and maps of early explorers and the evidence of modern

archaeological fieldwork. Useful secondary source books with maps
include William Sturtevant's *Handbook of North American Indians:
Southwest,* Edward Spicer's *Cycles of Conquest* and Donald Meinig's
Southwest: Three Peoples in Geographic Change, 1600-1970.[1]

The "Spanish Conquest" 1540-1820 map depicts a new geography of the Southwest wrought by the Spanish who initially followed
Indian trails and then built cart roads north from Mexico City
(Map 27). These alien intruders were a potent force for geographic
change — taking lands, establishing presidios and missions, redi-
recting Indian labor and economic activity, opening mines and
mining towns, introducing domesticated animals, and making maps
with new names on the land. The settled "Spanish" Southwest
showed a distinct European influence, yet the larger region retained
elements of and ties to the earlier Indian presence. Primary source
material was used in the selection, location and proper spelling of
Spanish and Mexican places shown on the map and included the
following sources:

Passage Par Terre a la California[2]
By Francois Kino
ca. 1702

Nicolas de Laforas Description[3]
Mapa de la Frontera del Vireinato de Nueva Espana
1771

The Escalante Trail[4]
1776

Plano de la Provincia Interna de el Nuevo Mexico[5]
1779

Perhaps most important was the excellent reproduction of numerous
period maps found in Volume 1 of Carl Wheat's superb series,
Mapping the Transmississippi West. Also useful were the contemporary maps in W. P. Cumming's work, *The Exploration of North
America 1630-1776.*[6]

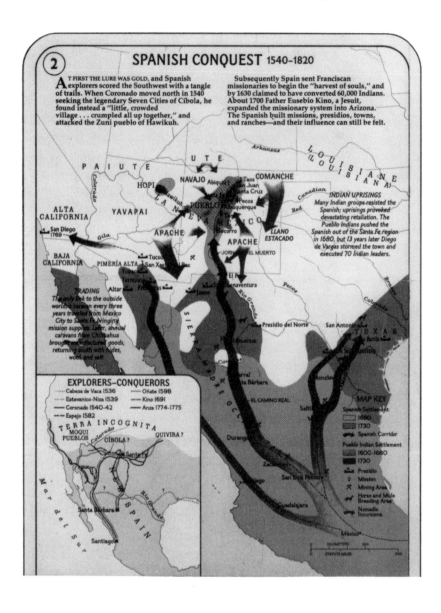

SPANISH CONQUEST 1540-1820

AT FIRST THE LURE WAS GOLD, and Spanish explorers scored the Southwest with a tangle of trails. When Coronado moved north in 1540 seeking the legendary Seven Cities of Cíbola, he found instead a "little, crowded village . . . crumpled all up together," and attacked the Zuni pueblo of Hawikuh.

Subsequently Spain sent Franciscan missionaries to begin the "harvest of souls," and by 1630 claimed to have converted 60,000 Indians. About 1700 Father Eusebio Kino, a Jesuit, expanded the missionary system into Arizona. The Spanish built missions, presidios, towns, and ranches—and their influence can still be felt.

INDIAN UPRISINGS
Many Indian groups resisted the Spanish; uprisings provoked devastating retaliation. The Pueblo Indians pushed the Spanish out of the Santa Fe region in 1680, but 13 years later Diego de Vargas stormed the town and executed 70 Indian leaders.

TRADING
The only link to the outside world's caravan every three years traveled from Mexico City to Santa Fe, bringing mission supplies. Later, annual caravans from Chihuahua brought manufactured goods, returning south with hides, wool, and salt.

EXPLORERS–CONQUERORS
- Cabeza de Vaca 1536
- Estevanico-Niza 1539
- Coronado 1540-42
- Espejo 1582
- Oñate 1598
- Kino 1691
- Anza 1774-1775

MAP KEY
Spanish Settlement
- 1680
- 1730
- Spanish Corridor
Pueblo Indian Settlement
- 1600-1680
- 1730
- Presidio
- Mission
- Mining Area
- Horse and Mule Breeding Area
- Nomadic Incursions

Fig. 27.

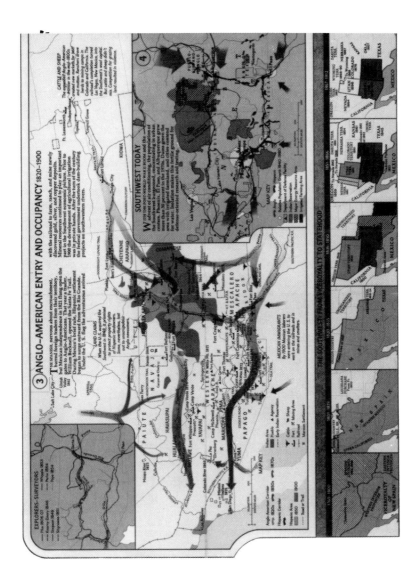

Fig. 28.

The "Anglo-American Entry and Occupancy" 1820-1900 map portrays the early period of the last major occupance of the Southwest that continues to the present (Map 28). The Americans came to the Southwest in three discernible periods and from three directions: first beginning in the 1820s the traders and soldiers following the Santa Fe Trail from Missouri, then in the 1850s the cattlemen from Texas, and lastly in the 1870s the missionaries, miners and farmers from the East and Mormon settlers from Utah. The landscapes that the Americans found were reshaped by them to reflect new economies, settlements, networks of transportation, and a new organization of the land. The Indian population was removed to government reservations while wagon roads were supplemented by railroads. Anglo towns with rectangular street patterns and central squares began to emerge along the railroads, old resource areas were abandoned and new areas opened. The power authority of the U.S. government was ever present at army posts.

Primary map sources used included an 1846 map of Texas, Oregon and California and the regions adjoining, an 1855 map of routes for a Pacific railroad, an 1875 map of the District of New Mexico, maps of the territory of New Mexico and Arizona prepared by the General Land Office in 1903, and numerous reproductions of period maps in Carl Wheat's *Mapping the Transmississippi West,* (volumes 2 through 5).[7]

The "Southwest Today" map shows salient aspects of the contemporary geography of the region. New landscapes have evolved that include large urban centers, irrigated agriculture, interstate highways, an enlarged military presence, exploitation of major energy resources, and the arrival of people from northern states and from Mexico. While the map reflects the mid 20th century, it also mirrors remnants of cultural landscapes that existed previously. The map sequence "The Southwest From Viceroyalty to Statehood" 1776-1912 summarizes the long and complex political geography of the Southwest.

The large "Southwest" map on the other side of the sheet displays the contemporary scene showing the physical and cultural diversity of Arizona and New Mexico and parts of the surrounding states that make up the "Southwest" region. Many historic sites that are accessible to the traveler are shown.

The actual production of the Southwest map, — from concept to completion, embodies state of the art cartography in terms of equipment and materials, the work of a highly professional staff of some sixty people, and printing 11 million copies for an international audience.

Initially we had to define the scope and content of the "Making of America" series. This challenge meant setting forth with the aid of our chief editorial consultant, Dr. D.W. Meinig, Maxwell Professor of Geography at Syracuse University, the essence of the American experience in map form attractive to and readily understandable by a broad readership. Our purpose is to provide a fresh perspective on the continuous and dynamic development of the United States emphasizing a geographic interpretation of our history and focusing on how each region owes its special character to the changing human geography of its past and how the imprint of former inhabitants is often still there to be read in the landscape. Each map has a principal regional consultant. Dr. Richard Nostrand, Professor of Geography at the University of Oklahoma, served in this capacity for the Southwest map. Other consultants are called upon to provide special topical or regional expertise.

The first step in the preparation of the Southwest map was the design phase where our map designers determine how best to present the historical geography of the Southwest on a single large sheet of paper. Concerns of designers include the individual inset map sizes reflecting the relative importance of the historical periods, map colors, symbols, and patterns used to convey historical geography information, the balance of art work, text, and inset maps on the map sheet, and of course the visual design continuity of the entire "Making of America" series.

The second step in the preparation of the map focuses on what the map tells the reader, or in other words, map content. This is the job of researchers, consultants, writers, and compilers. For the Southwest map, researchers gathered and analyzed hundreds of pieces of primary and secondary source material. Ten consultants were used and their input was reconciled into the project. The text writer undertook independent research to prepare the more than 100 notes in many different versions. The compilers began preparing bases and concept sketches and then the final compilation reflecting exact specifications for type, symbols, and patterns.

The third phase in preparing a map is actual production — the mechanical work of putting all the pieces together. Draftspersons scribe lines for rivers, shorelines, boundaries, railroads, roads, and mark town spots. Typographers place type on prepared base maps. The photomechanical laboratory then produces color negatives for the preparation of printing plates. With final negatives and printing plates prepared, map proofs are run so that the detailed and extensive editorial process can begin, to make sure that all information is correct and presented in a readable and attractive manner.

After final release of the map by the editors the printing process begins at a commercial printing company. The printing of 11 million copies on a six color web press takes approximately three weeks of nonstop high speed printing. Folding and trimming follows and the maps are then transported in trucks to W.F. Hall Company in Corinth, Mississippi to be inserted into the magazine. To date, nine maps of "The Making of America" series have been completed: *The Southwest* (November 1982), *Atlantic Gateways* (March 1983), *The Deep South* (August 1983), *Hawaii* (November 1983), *Alaska* (January 1984), *Far West* (April 1984), *Central Rockies* (August 1984), *Northern Approaches* (February 1985), and *Central Plains* (September 1985). The remaining maps of the series will be completed by 1988. These regions are: Tidewater, New England, Ohio Valley, Great Lakes, Northern Plains, Texas, Pacific Northwest, and the West Indies. The series is being produced in anticipation of the centennial celebration of the Society in 1988. In commemoration of the centennial it is anticipated that "The Making of America" series will be recast and expanded with additional maps, photographs and text into an Atlas of American History befitting the event and the Society's tradition of excellence.

NOTES

1. William C. Sturtevant, ed., *Handbook of North American Indians, Vol.* 10; Southwest (Washington, D.C.: Smithsonian Institution, Bureau of Ethnology, 1978); Edward H. Spicer, *Cycles of Conquest: The Impact of Spain, Mexico, and the United States on the Indians of the Southwest, 1553-1960* (Tucson: The University of Arizona Press, 1962); D.W. Meinig, *Southwest: Three Peoples in Geographical Change,* 1600-1970 (New York: Oxford University Press, 1971).

2. Map of *Passage Par Terre a La California,* ca. 1702, by Father Eusebio Francisco Kino and reproduced in W.P. Cumming, S.E. Hillier, D.B. Quinn, and G. Williams, *The Exploration of North America 1630-1776* (New York: G.P. Putnam's Sons, 1974), p. 236.

3. *Mapa de la Frontera del Vireinato de Nueva Espana,* 1771, Nicholas de Lafora and reproduced in Lawrence Kinnaird, *The Frontiers of New Spain: Nicholas de Lafora's Description 1766-1768* (Berkeley: The Quivira Society, 1958).

4. *Plano Geographico de la Tierra Descubierta, y Demarcada, por Dn Bernardo de Miera y Pacheco al rumbo del Noreste, y Oeste de Nuevo Mexico . . . . ,* 1777, by Bernardo Miera y Pacheco. Manuscript size 32½ by 27¾ inches. Photostat copy in Library of Congress.

5. *Plano de la Provincia Interna de el Nuevo Mexico . . . ,* 1779, and reproduced in *The Missions of New Mexico, 1776: A Description by Fray Francisco Atanasio Dominguez with other Contemporary Documents.* Translated and annotated Eleanor B. Adams and Fray Angelico Chavez (Albuquerque: The University of New Mexico Press, 1956), p. 1269.

6. Carl I. Wheat, *Mapping the Transmississippi West,* 5 Vols. (San Francisco: The Institute of Historical Cartography, 1957-63); Cumming, *et, al, The Exploration of North America.*

7. *A New Map of Texas, Oregon, & California with the Regions Adjoining,* 1846, by Samuel A. Mitchell, and reproduced in Wheat, *Mapping the Transmississippi West,* Vol. 3, opp. p. 29; *Map of Routes for a Pacific Railroad,* 1855, compiled to accom-

pany the Report of the Hon. Jefferson Davis, Sec. of War, and reproduced in Wheat, *Mapping the Transmississippi West*, Vol. 4, opp. p. 24; map of the *District of New Mexico*, 1875, prepared by Lt. C. C. Morrison, U.S. army, scale 1:1,200,000, copy in Library of Congress; map of the *Territory of New Mexico*, 1903, prepared by the General Land Office, Washington, D.C. scale 1:790,320, copy in Library of Congress.

This book is published for
The Map and Geography Round Table
of the
American Library Association
by
Speculum Orbis Press
207 W. Superior
Chicago, Illinois 60610

The Paper is Litho mat
The type is Garamond No. 2

Book design by Donna P. Koepp
with Kathy Snyman

Typesetting donated by
Wesley A. Brown
and
Bowne of Denver

Printed by
Edwards Bros. Inc.
Ann Arbor, Michigan 48106